# ROUTLEDGE LIBRARY EDITIONS: SOCIAL AND CULTURAL GEOGRAPHY

Volume 12

# DAVID HARVEY'S GEOGRAPHY

# DAVID HARVEY'S GEOGRAPHY

JOHN L. PATERSON

Routledge
Taylor & Francis Group

LONDON AND NEW YORK

First published in 1984

This edition first published in 2014
by Routledge
2 Park Square, Milton Park, Abingdon, Oxfordshire OX14 4RN

and by Routledge
711 Third Avenue, New York, NY 10017

*Routledge is an imprint of the Taylor and Francis Group, an informa business*

First issued in paperback 2015

*British Library Cataloguing in Publication Data*
A catalogue record for this book is available from the British Library

ISBN 978-0-415-83447-6 (Set)
eISBN 978-1-315- 84860-0 (Set)
ISBN 978-0-415-73348-9 (hbk) (Volume 12)
ISBN 978-1-138-99060-9 (pbk) (Volume 12)
ISBN 978-1-315-84832-7 (ebk) (Volume 12)

**Publisher's Note**
The publisher has gone to great lengths to ensure the quality of this reprint but points out that some imperfections in the original copies may be apparent.

**Disclaimer**
The publisher has made every effort to trace copyright holders and would welcome correspondence from those they have been unable to trace.

# David Harvey's Geography

John L. Paterson

CROOM HELM
London & Canberra
BARNES & NOBLE BOOKS
Totowa, New Jersey

© 1984 J.L. Paterson
Croom Helm Ltd, Provident House, Burrell Row,
Beckenham, Kent BR3 1AT

British Library Cataloguing in Publication Data

Paterson, John
    David Harvey's geography.
    1. Harvey, David, 1935 Oct. 31  2. Geography — Philosophy
    I. Title
    910'.01   G96
    ISBN 0-7099-2029-6


First published in the USA 1984 by
Barnes and Noble Books
81 Adams Drive
Totowa, New Jersey, 07512

ISBN 0-389-20441-2

Printed and bound in Great Britain

# CONTENTS

# LIST OF FIGURES

# PREFACE

This book is a revised version of a thesis submitted for the Master of Philosophy degree at the University of Waikato, Hamilton, New Zealand, in December 1980. Not only has the original work been updated and edited, but also significant sections of the study, especially the second half, have been rewritten. Unfortunately, Harvey's latest book, *The Limits to Capital* (Basil Blackwell, Oxford, 1982), was not available to me until after the completion of the revised manuscript. However, in that book Harvey confirms the analysis that I present in Chapters 4 and 5 regarding the existence of a fundamental tension in his Marxist writings. This tension gives rise to a separation between theoretical and concrete historical analyses and the consequent problem of successfully integrating the two. In Harvey's own words,

> I think it is possible to . . . transcend the seeming boundaries between theory, abstractly formulated, and history, concretely recorded; between the conceptual clarity of theory and the seemingly endless muddles of political practice. But time and space force me to write down the theory as an abstract conception, without reference to the history. In this sense the present work is, I fear, but a pale apology for a magnificent conception. And a violation of the ideals of historical materialism to boot. In self-defence I have to say that no one else seems to have found a way to integrate theory and history (*The Limits to Capital*, 1982, p. xiv).

I originally undertook this research on Harvey's writings in order to gain some familiarity with two major philosophical influences in contemporary human geography, namely logical positivism and Marxism, and to explore the relationships between philosophy, methodology and geographical research. The emphasis in the book is therefore upon these issues rather than upon a biography of David Harvey. If some of my more critical comments prove eventually to have been misguided or unfounded, then I hope that a contribution nevertheless will have been made to the understanding of one of the most innovative and iconoclastic scholars in contemporary Anglo-American human geography.

J.L.P., Vancouver.

## ACKNOWLEDGEMENTS

The research upon which this book is based was undertaken in the Department of Geography at the University of Waikato, Hamilton, New Zealand, in 1979 and 1980. I would like to acknowledge the guidance, assistance and inspiration of teachers and friends among the staff and students of that Department. Professor Craig Duncan, my chief supervisor, actively encouraged and assisted me throughout the original project. Both Ann Magee, my second supervisor, and Lex Chalmers provided moral support and inspiration at crucial times. Lex has also rendered invaluable practical assistance. When I was employed in the Registry of the same University, I received some very practical assistance with the task of revision from friends and colleagues there. Another friend, Fletcher Cole, undertook the majority of the work on the index. Above all, I must record my thanks to my wife, Jean, and to my parents, who have unceasingly supported me in many ways.

I would like to thank Edward Arnold Ltd, London, for permission to reproduce extracts from Harvey's *Explanation in Geography* (1969) and *Social Justice and the City* (1973), and the Institute of British Geographers for permission to reproduce Figures 1 and 2 from R.T. Harrison and D.N. Livingstone, 'Philosophy and Problems in Human Geography: a Presuppositional Approach', *Area*, 12, 1980, 25-31.

Finally, I would like to record my debt to the late S.S. Cameron who, in 1976, at the University of Otago, Dunedin, New Zealand, guided me in my initial studies in the history and philosophy of geography. This book is dedicated to his memory and to the advancement of a genuine philosophical pluralism in geography.

# DAVID HARVEY

1961 'I have no intention of trying to erect any elaborate mansion of theory around the conclusions already established only to find that we are still obliged "to live in the shack outside"' (1961, p. ix)

1967 'The student of . . . geography . . . can either bury his head, ostrich-like, in the sand grains of an idiographic human history, conducted over unique geographic space . . . Or he can become a scientist . . . The first stage of scientific investigation involves the careful testing of theories and generalization, against the facts' (1967a, p. 551)

1969 '"The quest for an explanation", writes Zetterberg (1965, p. 11), "is a quest for theory" . . . Without theory we cannot hope for a controlled, consistent, and rational, explanation of events. Without theory we can scarcely claim to know our own identity' (1969a, pp. 89, 486)

'Geographers . . . [have been] failing, by and large, to take advantage of the fantastic power of the scientific method' (1969a, p. vi)

1970 'Shortly after [coming to America], I happened to read Karl Marx . . . The more I read, the more it seemed to make sense . . . We could, Marx suggested, create a theory to explain the contradictions [of capitalism] at the same time as it would help us to overcome them. I found this very exciting and started to work at it. And lo and behold, one by one, the contradictions . . . crumbled before the power of the analysis' (1978e, p. K2)

1972 'There is a clear disparity between the sophisticated theoretical and methodological framework which we are using and our ability to say anything really meaningful about events as they unfold around us . . . The most fruitful strategy at this juncture is to explore that area of understanding in which certain aspects of positivism, materialism and phenomenology overlap to provide adequate interpretations of the social reality in which we find ourselves. This overlap is most clearly explored in Marxist thought' (1972c, pp. 10-11)

1973　'A genuinely humanizing urbanism has yet to be brought into being. It remains for revolutionary theory to chart the path from an urbanism based in exploitation to an urbanism appropriate for the human species. And it remains for revolutionary practice to accomplish such a transformation' (1973a, p. 314)

1975　'The most important revelation I gained from reading Marx is that there is much mystification around . . . We have to begin a process of demystification and build a critical theory which will reveal to us what is really happening . . . In order to change the world we have to understand it' (1975b, p. 55)

1978　'The struggle between [Marxism and bourgeois social science] is a struggle to establish a hegemonic system of concepts, categories and relationships for understanding the world. It is a struggle for language and meaning itself' (1978c, p. 28)

　　　'My objective is to understand the urban process under capitalism . . . I hang my interpretation . . . on the twin themes of *accumulation* and *class struggle*' (1978b, p. 101)

# 1 INTRODUCTION

## David Harvey's Geography, 1961-1981

'I am at a disadvantage in discussing, with Stephen Gale, *Explanation in Geography*, because I have never read it' (Harvey, 1972f, p. 323).

'The important question for me is where I am going, not where I have been' (Harvey, personal communication, October, 1981).

Given the present state of philosophical and methodological pluralism in geography (Bird, 1979) it is important that human geography becomes a more self-critical and reflective discipline. Thus, while Harvey presses forward in his scholarship, it is necessary to begin an assessment of the past and present work of this influential scholar. This book is an analysis of the thought of David Harvey as it is expressed in his geographical publications over the period from 1961 to 1981. The philosophical and methodological aspects of Harvey's work will be emphasised, one of the main concerns of the study being the relationship between philosophy, methodology and geography as apparent in Harvey's writings.

Harvey was awarded a Bachelor of Arts degree in geography at the University of Cambridge, England, in 1957.[1] In the following years, he wrote a doctoral thesis entitled 'Aspects of Agricultural and Rural Change in Kent, 1800-1900', which was completed in October 1960, and 'approved by the Board of Research Studies' in January 1961 (Harvey, 1961). Harvey then took up an assistant lectureship in geography at the University of Bristol, later a full lectureship, and remained at Bristol until 1969 (Peel, 1975, p. 415, fig. 20.1), when he moved to Johns Hopkins University, Baltimore, Maryland, in the United States of America. There he took up a professorship in the Department of Geography and Environmental Engineering. At the time of writing, Harvey is still at Johns Hopkins University.

In 1963, Harvey published an article based on his doctoral research, and then, between 1966 and 1969, published eight articles, all dealing with methodological and theoretical questions in geography. In 1969, Harvey's first book was published. Entitled *Explanation in Geography*, it was partly based upon about five years of teaching an undergraduate

1

course on methodology in geography at the University of Bristol (Harvey, 1969a, pp. v-vi). Johnston, in his book on recent Anglo-American human geography, described *Explanation in Geography* as 'the first major work on the philosophy of the "new geography" . . ., a book which received wide acclaim' (Johnston, 1979, p. 62). A physical geographer, in a contemporary review in the *British Journal for the Philosophy of Science*, hailed Harvey's book as 'the most authoritative and helpful statement of "The Nature of Geography" since the publication of Richard Hartshorne's classic of that title in 1939' (Kennedy, 1970, p. 401). Later, in an article presenting a Marxist interpretation of the 1960s and 1970s in geography, Peet (1977b, p. 249) called it 'the bible of new (theoretical) geography'.

After moving to Baltimore, Harvey became involved in research on that city and published four articles on conceptual problems iñ geography, centring on the relationship between social processes and spatial form, and exploring how notions like 'social justice' might be incorporated into geographical research. He gathered together these articles, added two previously unpublished papers, and in 1973 had them published under the title of *Social Justice and the City*. Here, Harvey presented the 'convergence' of his thought with that of Karl Marx (Harvey, 1973a, pp. 287, 301). Once again, contemporary reviewers saw the book as of central importance to the discipline. Hall, a British urban geographer, stated that 'this without doubt is one of the most significant contributions to geographical thought to emerge in the last two decades' (Hall, 1973, p. 409). Cox, who had moved to the United States after studying at the University of Cambridge, and who had written an undergraduate text on human geography (Cox, 1972), called *Social Justice and the City* 'a provocative and multifaceted book likely to have an enduring impact upon human geography in general and urban studies in particular' (Cox, 1976, p. 333). A French Marxist urban sociologist has referred to Harvey's book and his subsequent work as 'exemplary Marxist work . . ., still an exception [in the United States]' (Castells, 1977, p. 470).

Over 1973 and 1974, Harvey published three papers on Marx's method and then concentrated upon utilising that method in the study of urbanisation in advanced capitalist countries. He has published eleven articles on this subject and, since 1975, has made reference to a forthcoming book on 'urbanization under capitalism' (Harvey, 1975d, p. 9, note), a book 'that seems to take an interminable time to finish' (Harvey, 1976b, p. 80) but 'which may see the light of day shortly' (Harvey, 1978b, p. 130). In fact, Harvey's third book, *The Limits to Capital*, was eventually published in 1982. The most recent of Harvey's published articles, 'Monument and Myth', was based on research in Paris in 1976-1977 and dealt with the symbolical and mythical meanings of the Basilica of Sacré-Coeur (Harvey, 1979). He has also written a number of short newspaper and magazine articles. A comprehensive bibliography of Harvey's writings is contained in an appendix to this book.

David Harvey has been a prominent international contributor to the philosophical and methodological debates within Anglo-American geography over the last two decades. In the summer of 1964, for instance, Harvey attended a conference on spatial statistics at Northwestern University, Illinois, while he spent 1965-1966 at the Pennsylvania State University and made a short visit to the Australian National University, Canberra, in 1968 (Harvey, 1969a, p. ix; Brookfield, 1973, p. 8, note). Harvey has presented papers regularly at the annual conferences of the Institute for British Geographers, for instance in 1973, 1974 and 1977, as well as at those of the Association of American Geographers, for instance in 1980 and 1981.

Harvey's methodological work in the 1960s, especially his *Explanation in Geography*, has been characterised as 'logical positivist', although to varying degrees. Guelke (1978, p. 35) saw *Explanation in Geography* as 'a thorough logical positivist analysis of geographic explanation'; Mercer and Powell (1972, p. 38) commented that Harvey demonstrated a 'preference' for logical positivism; and both Chisholm (1975, pp. 124-5) and Gregory (1978, pp. 33-4) noted Harvey's endorsement of Hempel's explanatory model, which is central to the logical positivist view of science.

In Chapter 1 of *Explanation in Geography*, Harvey discussed 'Philosophy and methodology in geography' and made the following observations:

> There are some philosophers, logical positivists of the extreme variety, who have held that all knowledge and understanding can be developed independently of philosophical presuppositions. Such a view is not now generally held, for logical positivism in such an extreme form has turned out to be barren. Methodology without philosophy is thus meaningless. Our ultimate view of geography must therefore take both methodology and philosophy into account. Such an ambitious synthesis will not be attempted here, for before we can hope to achieve it, we need a much better understanding of methodological problems alone. But although the emphasis in this book is primarily upon methodological problems, we will have cause

on several occasions to refer to important philosophical issues concerning the nature of geography (Harvey, 1969a, p. 8).

It is not clear from Harvey's remarks as to the philosophical position
from which he was writing. Was he rejecting only the extreme form of
logical positivism or logical positivism itself? What was Harvey's philosophical stance in *Explanation in Geography*, or did he succeed in
excluding philosophy from the main considerations of the book? These
questions will be addressed particularly in Chapter 2 of this study.

After writing *Explanation in Geography*, Harvey 'began to explore
certain philosophical issues which had deliberately been neglected in
that book' (Harvey, 1973a, p. 9). He examined how ideas in social and
moral philosophy might be related to such topics as urbanism and urban
planning. The six papers brought together in *Social Justice and the
City* represent the evolution of Harvey's thought as he conducted this
exploration. Along the way, he became increasingly aware of the inadequacies of the 'positivist basis of the 1960s' in geography and he
began to consider seriously Marxist theory 'in which certain aspects of
positivism, materialism and phenomenology overlap' (Harvey, 1973a,
p. 129). Harvey pointed out that Marx had developed a phenomenological basis in his early writings and that both Marxism and positivism
had a 'materialist base' and 'analytic method'.

> The essential difference, of course, is that positivism simply seeks
> to understand the world whereas Marxism seeks to change it. Put
> another way, positivism draws its categories and concepts from
> an existing reality with all its defects while Marxist categories and
> concepts are formulated through the application of the dialectical
> method to history as it unfolds, here and now, through events and
> actions. The positivist method involves, for example, the applica-
> tion of traditional bi-valued Aristotelian logic to test hypotheses . . .
> Hypotheses are either true or false and once categorized remain ever
> so. The dialectic, on the other hand, proposes a process of under-
> standing which allows the interpenetration of opposites, incorporates
> contradictions and paradoxes, and points to the processes of resolu-
> tion . . . Truth lies in the dialectical process rather than in the state-
> ments derived from the process. These statements can be designated
> as "true" only at a given point in time and, in any case, can be contra-
> dicted by other "true" statements. The dialectical method allows us
> to invert analyses if necessary, to regard solutions as problems, to
> regard questions as solutions (Harvey, 1973a, pp. 129-30).

A question posed by Harvey's statements is: in rejecting the 'positivist method' based on 'traditional bi-valued Aristotelian logic', was Harvey rejecting Hempel's model of scientific explanation, which was central to his earlier writings (Harvey, 1967c, 1969a, pp. 36-41)? If Harvey had rejected Hempel's model, what then was the common 'analytic method' and 'materialist base' of positivism and Marxism to which he referred? What exactly was Harvey's view of Marx's method? And in his application of this method in his study of urbanisation, did Harvey modify it in any way? Duncan and Ley (1982) have argued that some of Harvey's Marxist writings take a 'structuralist' stance and are characterised by a holistic mode of explanation, in which reified entities such as 'capital' are treated as the formal cause whereas people are effectively regarded as mere carriers of a structural logic. On the other hand, other parts of Harvey's Marxist writings were seen by Duncan and Ley to be empirical studies that made few essential links with the theoretical framework of structural Marxism. Harvey was thus presented by them as an example of a fundamental dichotomy in Marxist thought between scientific Marxism, a structuralist tradition focusing on the theoretical treatment of political economy, and critical Marxism, a humanist tradition focusing on concrete historiographic study. Harvey (1973a, p. 288) regarded Marx's view of society as based upon 'operational structuralism' but he vigorously denied that it was deterministic and emphasised that 'Marx was a humanist' (Harvey, 1973b, pp. 32, 17). This constellation of issues will be addressed particularly in Chapters 3 and 4 of this study.

The analysis of Harvey's writings in Chapters 2, 3 and 4 attempts to clarify Harvey's philosophical positions, in relation to both his so-called 'logical positivist' works (Chapter 2) and his so-called 'Marxist' works (Chapters 3 and 4). When focusing upon the transition from one to the other (Chapter 3), an attempt will be made to isolate Harvey's dissatisfactions with his 'logical positivist' works, and to outline what he saw to be the strengths of a 'Marxist' approach.

This study centres on the philosophical and methodological aspects of Harvey's work during the last two decades. It thus falls within the fields of the philosophy and history of geography. Following a review of recent writings in the philosophy of geography, the literature in the history of science and of geography relevant to a study of philosophical and methodological change is reviewed. In the final section of this introductory chapter, consideration is given to the principles of historical philosophical study in geography.

**Philosophy, Method and Geographical Research**

The following account of philosophical discussion in Anglo-American, mainly human, geography since the 1930s is only introductory and will be expanded upon at the beginnings of Chapters 2, 3 and 4.

Between about 1930 and 1960, philosophical discussion within geography largely involved consideration of the nature, scope and objects of geographical study and the relationship between geography and other disciplines (for example, Roxby, 1930; Darby, 1953; Philbrick, 1957). Hartshorne's (1939) *The Nature of Geography: A Critical Survey of Current Thought in the Light of the Past* both gave expression to and further shaped the intellectual milieu of Anglo-American geography during this period. Emphasising 'areal differentiation' and the regional character of geographical study, Hartshorne's work was seen as advocating an 'idiographic' method that eschewed the development of laws in geography (Gregory, 1978, pp. 30-1). During the 1950s, a 'spatial science' approach was developed, which came into prominence in the 1960s. Debate continued on the nature, scope and objects of geography in the light of the then new approach (for example, Ackerman, 1963; Berry, 1964a; Brookfield, 1964; Ackerman *et al.*, 1965), but there was a noticeable tendency for 'geographical philosophizing', as Harrison and Livingstone have put it, to become restricted to 'the internal structure of explanation' (Harrison and Livingstone, 1980, p. 25). As already noted, it has been argued that Harvey's (1969a) *Explanation in Geography* stood in relation to the philosophical and methodological discussion of spatial science geography as Hartshorne's (1939) work stood in relation to that of the earlier regional geography (Kennedy, 1970, p. 401; Johnston, 1979, p. 62).

The first published criticism of spatial science geography came from those geographers who were dissatisfied with what they considered to be its often implicit neo-classical economic assumptions. Throughout the late 1960s, such geographers advocated a behavioural approach, emphasising aggregative spatial behaviour and decision-making. The early 1970s witnessed a growing concern within geography for 'relevant' research and teaching. For instance, between 1971 and 1975, there was a debate published in *Area*, a journal of the Institute of British Geographers, over the contribution of the discipline to the solution of urgent social problems. Debates over the 'internal structure of explanation', the behavioural approach and the question of relevance in geography were proceeding at the same time as a radical political and philosophical perspective was being developed in North American human geography

(Smith, 1971). Indicative of this trend was the founding of *Antipode*, subtitled 'A Radical Journal of Geography', by a group of staff and students at Clark University, Massachusetts, in 1969. Peet later saw Harvey's (1972d) article, 'Revolutionary and Counter-revolutionary Theory in Geography and the Problem of Ghetto Formation', as finally leading 'the breakthrough from liberal to Marxist geography' (Peet, 1977b, pp. 249-50).

In 1970, criticism of spatial science geography was introduced from another philosophical quarter, that of phenomenology (Relph, 1970). Philosophical debate in geography thus began to take account of different philosophical systems instead of dealing mainly with methodology. The range of perspectives considered in the geographic literature since the early 1970s has grown to include a broader 'humanistic' approach (Tuan, 1976; Ley and Samuels, 1978), which incorporates phenomenology, existentialism (Samuels, 1978) and a neo-Kantian perspective (Berdoulay, 1976; Entrikin, 1977). In 1974, Guelke proposed the alternative of idealism, and Olsson (1974, 1975, 1980), in exploring the 'utopian optimism' of Marx and the 'realistic pessimism' of Wittgenstein, has developed what might well be called a 'dialectical idealism' (see also Marchand, 1974, 1978). Thus, in the 1970s, Anglo-American human geography moved into a period of philosophical and methodological pluralism which remains the intellectual climate of the discipline today.

The last decade of philosophical debate in geography might be characterised as, to use a phrase from another context, 'a pursuit of truth as distinct from the pursuit of technically reliable knowledge' (Kolakowski, 1975, p. 7). Whereas the 1960s were largely concerned with one particular methodology, the debates of the 1970s have incorporated other methodologies as well as a wide range of different philosophical issues, centring on the meaning of science, its relation to political, economic and social life, the role of ethical notions in geographical theory, the limits of quantification, the origin of knowledge, the possibility of objectivity and the extent of subjectivity. What is at stake seems to be, on the one hand, certitude, and on the other, meaning.

Of course, it is always possible for a geographer to believe that philosophical considerations have a very limited influence upon geographical teaching and research. However, an inescapable interrelationship between philosophy, method and geographical research has recently been firmly asserted by a number of geographers, for instance Sayer (1979), Harrison and Livingstone (1980), Harvey and Holly (1981b) and Hill (1981). This has been an area of concern to the present author as well (Paterson, 1976, 1977). Sayer attempted to develop a materialist view of unstated and implicit assumptions in geography. He concluded that

> a lack of awareness of the influence of theoretical, metaphysical and
> epistemological presuppositions in our studies of what we believe to
> be matters of fact means that our geographical knowledge is grounded
> not in the firm bedrock of such "facts" but in the uncorrected, un-
> examined and frequently incoherent presuppositions of vulgar com-
> mon sense (Sayer, 1979, p. 38).

Harvey and Holly introduced a collection of essays on *Themes in Geographic Thought* by asserting that 'geographical thought, at any point
in time, is a manifestation of the interaction between the prevailing
philosophical viewpoints and the major methodological approaches in
vogue' (Harvey and Holly, 1981b, p. 11). They went on to suggest a
generalised pattern of relationships between methodology, philosophy,
theory and paradigms. Hill's essay in the same book discussed positivism
as a 'hidden philosophy' in geography. Noting the fact that many geographers doubt that philosophical issues are actually relevant to their
research, Hill nevertheless asserted that

> Even if it is not explicitly articulated, all research is guided by a set
> of philosophical beliefs . . . [which] influence or motivate the selec-
> tion of topics for research, the selection of methods for research,
> and the manner in which completed research projects are subjected
> to evaluation. In short, philosophical issues permeate every research
> decision in geography (Hill, 1981, p. 38).

Harrison and Livingstone have constructed the most comprehensive
conceptual framework to date for viewing 'the pervasive influence of
presuppositions in all scientific and philosophical thought' with special
reference to geography (Harrison and Livingstone, 1980, p. 25). There
are two parts to their framework. The first is a 'presuppositional hierarchy' (Figure 1.1), which aims 'to explore the origin and implications
of the tacit first premises' of every scientific investigation. Even the
specification of what is to be studied implies 'cosmological assumptions
about the nature and origin of reality'. Epistemological views, which re-
strict the kinds of methodological techniques thought to be relevant to
an investigation, are themselves seen to be 'generated' by cosmological,
ontological and disciplinary presuppositions (Harrison and Livingstone,
1980, p. 26). Such a presuppositional hierarchy shapes the conceptual
categories that are used in any geographical investigation.

Figure 1.1: The Presuppositional Hierarchy (Harrison and Livingstone, 1980, 27, fig. 1)

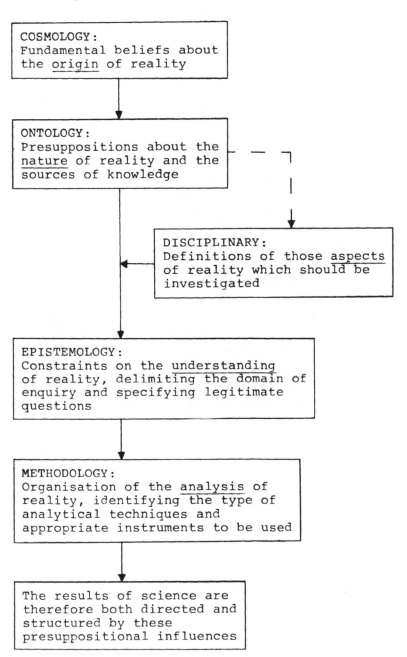

The second part of Harrison and Livingstone's framework consists of a 'problem cycle' (Figure 1.2). Having stated their commitment to 'changing the world' through an interdisciplinary approach to social problems, Harrison and Livingstone believe that a situation is perceived to be a problem only in the light of the investigator's presuppositions and subsequent orientations. The problem is formulated and then evaluated as to its significance. For the investigator to conclude eventually that a solution to the problem has been found involves an evaluation just as much as does the decision that the problem exists. And,

> recognising that the perception, formulation, evaluation and solution of problems are stages in an attempted reorientation of social patterns to some desired future state, any proposed solution is more a reflection of the way of looking at reality than of reality itself (Harrison and Livingstone, 1980, p. 29).

The political implementation of solutions creates another problem to be solved. Presuppositions are seen by Harrison and Livingstone to influence every stage in the problem cycle.

Harrison and Livingstone's views on the presuppositional character of geographical research potentially consist of a very fruitful approach to analysing the relationship between philosophy, methodology and geography. Their preliminary investigation into a reassessment of the influence of Immanuel Kant on the development of geography has arisen out of the view that philosophy shapes geography at many points, not at just the most apparent ones (Livingstone and Harrison, 1981). This long-overdue reassessment therefore emphasised the implications of Kantian philosophy generally for the development of geography. However, the potential of the presuppositional hierarchy and problem cycle was not developed in this article and the framework has yet to be grounded with reference to actual geographical studies. A number of matters regarding the nature of presuppositions therefore remain unclear. For instance, do geographers consciously accept certain presuppositions from which they will argue? If so, are such presuppositions susceptible to empirical verification? Or will any research that geographers conduct always implicitly assume a way of viewing reality, a view often not consciously recognised until it is reflected upon, which might be characterised as a constellation of philosophical assumptions or presuppositions? In the concluding chapter of this study, the opportunity will be taken to assess Harrison and Livingstone's views on the role of philosophical presuppositions in methodological and geographical research in the light of an analysis of Harvey's writings.

Figure 1.2: The Problem Cycle (Harrison and Livingstone, 1980, 29, fig. 2).

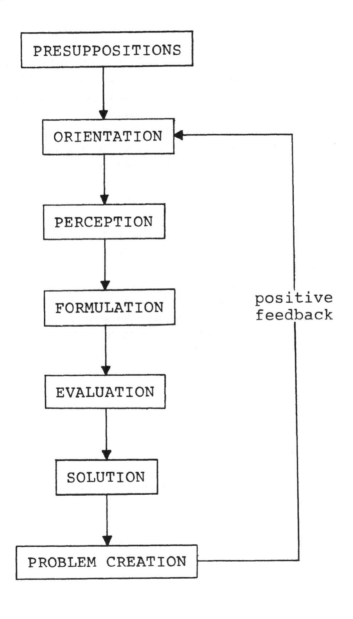

### The History of Science and the History of Geography

In a doctoral thesis entitled 'Conceptual Change and the Growth of Geographic Knowledge: a Critical Appraisal of the Historiography of Geography', Aay has noted that the progress of the discipline of the history of science over the last 30 years has left 'an extensive corpus of histioriographic debate of broad relevance on which the history of geography may draw with profit' (Aay, 1978, p. 84). 'Historiography' in Aay's thesis refers to those studies dealing with the methodological and philosophical standards of historical research. In the first half of the twentieth century, George Sarton helped to provide the history of science with a cognitive and institutional identity (Thackray and Merton, 1975). He presented the growth of science as an incremental chronicle of scientific achievements centred on the lives and works of great scientists. However, as Aay pointed out, his writings failed to provide 'the analytical depth necessary to understand the *content* of science and the *process* of scientific growth and discovery' (Aay, 1978, pp. 89-90).[2]

After World War II, alternative approaches to the history of science were developed. One such alternative emphasised the 'internal' process of scientific investigation, viewing the history of science as a history of ideas. Drawing its inspiration primarily from Alexandre Koyré (Kuhn, 1970d; Gillispie, 1973), this outlook aimed, by meticulous critical analysis of original documents, to examine the science of a period in its own terms. 'This would include mapping the intellectual milieu both generally, and as specifically related to the problem at hand, determining precisely the nature of the problem as seen by science or the scientist' (Aay, 1978, p. 91). Recent developments in the history of science, having their roots in the sociology of knowledge and of science and in Marxism and neo-Marxist thought, have sought to bring about a balance between the 'internal' and 'external' approaches. The progress of scientific knowledge is viewed as a product of 'a socio-economic base and cultural/ideological context, and [is] not ... internally and independently generated by rational thought and academic enterprises' (Aay, 1978. p. 124).

This study of Harvey's work is restricted to a consideration of his writings, presenting an 'internal history' of his thought as it is expressed in these writings. It is acknowledged that whereas an internal history is necessarily part of any study of a scholar's work, it is by no means the complete study itself. Any 'explanation' arising out of the analysis in this book of Harvey's turn to a Marxist perspective will therefore be, at best, only partial. For wholeness it must be supplemented by detailed biographical, institutional, social and other studies. As an internal history, this study has much in common with Koyré's approach. The internal historiography of science is currently represented by a number of schools of thought. At one extreme are Popper (1963, 1968, 1972a) and Lakatos (1970, 1971), who aimed to 'rationally reconstruct' the history of science 'in the light of current (positivist) standards in the philosophy of science' (Aay, 1978, p. 95). At the other extreme is T.S. Kuhn (1962, 1974), who admitted sociological considerations of scientific communities into his studies of conceptual change.

Among the recent literature on conceptual change in science, Kuhn's (1962) *The Structure of Scientific Revolutions* has had the most pronounced influence upon the manner in which geographers have both viewed and debated conceptual change in geography (Haggett and Chorley, 1967; Whitehand, 1970, 1971; Berry, 1973, 1978; Mercer, 1977 Bird, 1977, 1978; Buttimer, 1981). However, only Johnston (1979) has subjected Kuhn's model of scientific change to close scrutiny in the light of a detailed study in the history of geography. He concluded that Kuhn's analysis had only limited application and had to be supplemented by a branch model of the discipline within which generational change took place, conversion from one 'paradigm' to another being comparatively rare (Johnston, 1979, Chapters 1 and 7). This was perhaps to be expected, as Kuhn's concern was with the physical sciences and he did not consider his analysis necessarily relevant to the behavioural and social sciences (Aay, 1978, p. 8).

Central to Kuhn's viewpoint is the role of 'paradigms' in the progress of science. However, Masterman discovered 21 different senses in Kuhn's (1962) use of the term, although she saw these to fall into three main groups, 'metaphysical' paradigms or sets of beliefs, 'sociological' paradigms such as universally recognised scientific achievements, and 'artefact' paradigms such as actual textbooks (Masterman, 1970, pp. 61-5). In response to this, and to Shapere's (1964) and Buchdahl's (1965) criticisms, Kuhn (1970b, pp. 271-3, 1970c, pp. 181-91, 1974, pp. 460-72) has suggested that the term 'paradigm' should be replaced by two other terms, 'disciplinary matrix' (the sense of paradigm that geographers have taken up) and 'exemplar'. An exemplar is a standard example of a concrete problem and its solution, which may be applied to a range of other, as yet unsolved, scientific problems. A disciplinary matrix is composed of, among other things, a number of exemplars and sets of shared values, symbolic generalisations and models.

Toulmin (1970, 1972) has pointed out that the distinctive feature of

Kuhn's original account was not his theory of paradigms but his distinction between 'normal' and 'revolutionary' science. Normal science is conducted within the framework of a ruling disciplinary matrix. It tackles questions and problems that have meaning in relation to the disciplinary matrix, refining its research methods and measurement techniques and elaborating further elements of original achievements of the disciplinary matrix. Revolutionary science is that activity accompanying the revolutionary overthrow of one disciplinary matrix by another. The revolution is occasioned by the emergence of a number of serious anomalies unable to be dealt with satisfactorily by the ruling disciplinary matrix. For Kuhn, the progress of normal science is cumulative and unaccompanied by major theoretical debate. Revolutionary science, however, is innovative, and change from one disciplinary matrix to another is discontinuous, resulting in a new way of seeing the world (Kuhn, 1962, Chapter 10).

Between 1962 and 1965, Kuhn's critics questioned whether scientific change had ever been as revolutionary as Kuhn made out. He eventually conceded that theoretical change in science consisted of an unending sequence of smaller revolutions (Kuhn, 1970a). Toulmin (1972, p. 115) believed that, in fact, Kuhn had distinguished not two historical kinds of scientific change but rather two logically distinct aspects of any theoretical change in science.

Popper (1963) viewed the right to challenge scientific authority as one of the main features of genuine scientific activity. He regarded the history of science as a sequence of conjectures, refutations, revised conjectures and additional refutations (Losee, 1980, p. 171), as 'the repeated overthrow of scientific theories and their replacement by better or more satisfactory ones' (Popper, 1972b, p. 215). Theories were the basic unit of science for Popper, not disciplinary matrices and exemplars. 'We approach everything in the light of a preconceived theory' (Popper, 1970, p. 52), but science should never be dogmatic and uncritical.

> I do admit that at any moment we are prisoners caught in the framework of our theories; our expectations; our past experiences; our language . . . But . . . if we try, we can break out of our framework at any time. Admittedly, we shall find ourselves again in a framework, but it will be a better and roomier one; and we can at any moment break out of it again (Popper, 1970, p. 56).

Despite similar views on the revolutionary character of science, Kuhn and Popper hold quite different positions on the character of 'normal science'. The difference lies in the fact that Popper was judging all science in the light of his view of what science ought to be (Aay, 1978, p. 95). In the concluding chapter to this book, the opportunity will be taken to comment upon the applicability of both Kuhn's and Popper's accounts of conceptual change to such change within the work of one geographer.

The different approaches of Kuhn and Popper illustrate the influence of various philosophies of science in the study of the history of science (Lakatos, 1971, pp. 92-9). Logical positivism, inductivism, conventionalism and other philosophies of science all tend to reconstruct the history of science according to their view of the character of science. Aay suggested that models of scientific change may be 'tested against the weight of facts authenticated and interpreted by the history of geography' (Aay, 1978, p. 101). However, Stoddart (1981) has argued that geographers have distorted rather than clarified the history of geography when they have adopted Kuhn's terminology. The concept of revolution has been used to bolster the heroic self-image of those who viewed themselves as innovators, Kuhn's analysis being presented as an apparently scientific justification for the advocacy of change on social rather than strictly scientific grounds. Buttimer (1981) has pointed out that Kuhn's approach (and this applies also to Popper's) assumed that science is a style of thought separating mind and matter, subject and object, aiming at rational understanding and technical control over reality. She suggested an alternative approach to the history of geography which focused on the life work of individual geographers whose ideas have been significant in shaping the direction of geographic research in order to explore the reflective and personal dimensions of thought as well. Such an approach would seek to unmask the sociological and ideological influences on the connection between geographic ideas and praxis (Buttimer, 1981, p. 91).

This book is seen as a contribution towards a more reflective history of geography as well as a more self-critical praxis in contemporary geography. An attempt is made to represent Harvey's views faithfully, although the account is biased towards a consideration of the philosophical and methodological dimensions of his scholarship.

## Principles of Historical Philosophical Study

Aay has pointed out that there exist few historiographical guidelines to the history of geography within the geographical literature (Aay, 1978, p. 15). Discussion of the principles applicable to historical philosophical analysis and interpretation occurs even less frequently. As far as any initial analysis is concerned, Aay has pointed out that Koyré's approach is worthy of consideration.

> Koyré, in the history of science, focussed his attention on the philosophical examination of ideas. Here the essential matter is *explication du texte* – the careful and painstaking examination and comparison of original writings in order to grasp what the writer intended and meant, where his ideas came from and what new interpretations he added to them. Such textual analysis would need to range over the entire body of an individual scholar's writings or the literature of a research programme to fill out and particularize the meaning of concepts and to chronicle their changing meaning (Aay, 1978, p. 312).

Aay saw Van Paassen's (1957) study, *The Classical Tradition of Geography*, as 'one of the most consummate examples' of *explication du texte* in the history of geography (Aay, 1978, p. 322). By closely analysing the text of Harvey's writings and by interpreting them in accord with the principles of historical philosophical interpretation (considered immediately below), it ought to be possible to gain a sympathetic immanent appreciation of Harvey's views on their own terms. The outcome of such a detailed textual analysis, especially if it is placed against the background of the intellectual milieu of the time, provides philosophical discussion in geography with a concrete reference point. In this way, the relevance of philosophical considerations to geographical study may be demonstrated more clearly.

Only Hartshorne (1948), May (1972) and Livingstone (1979) have commented upon historical philosophical interpretation in geography. Hartshorne (1948, p. 116) called for the 'application of responsible scholarship' to the examination of 'methodological' writings in geography, which, for him, included philosophical writings. By responsible scholarship, Hartshorne meant 'scholarly discussion, based on critical examination of previous writings and cool logical analysis' (Hartshorne, 1948, p. 115). The epithet 'cool' was used by Hartshorne because one of the main concerns of his article was to assert that criticism of 'methodological' views did not imply personal criticism of the geographers who held those views (Hartshorne, 1948, pp. 121-5).

In reply to a critical review by Hartshorne of his study on *Kant's Concept of Geography* (1970), May considered two 'principles of historical philosophical interpretation' (May, 1972, p. 79). The first was that of 'coherence'.

> As one becomes more familiar with Kant's way of thinking, one becomes increasingly aware that certain statements cohere or fit in with that system of thought, whereas others do not. The basic assumption here is that a thinker of Kant's calibre is not going to blatantly contradict his own system of thought (May, 1972, p. 79).

However, Livingstone (1979, p. 228) has pointed out the pitfall of supplying a coherence which is otherwise lacking. Furthermore, May did not have to confront the problem of considerable development in a thinker's position. Inherent in the comparison of two works written by the same person, whether they be separated by a period of two weeks or two years, is the possibility of discontinuity in the stance or view presented. Even given a great deal of continuity in such writings, any 'system of thought' is bound to be dynamic, not static. To assume that Harvey 'will not blatantly contradict his own system of thought' has only limited application, given that Harvey appears prima facie to have changed his philosophical and methodological views considerably. What May has referred to as the principle of coherence has to be replaced by what will be called the principle of 'contextual interpretation', the principle of assuming that Harvey is to be interpreted initially on his own terms within the immediate context.

May's second principle of historical philosophical interpretation was that of 'imaginative reconstruction'.

> R.G. Collingwood once likened the historian to a detective; he finds any number of scattered clues and then attempts to weave them together to present a reasonable and integrated account of the matter. In the dialogue with Kant, one often finds no direct answers to the questions one poses. Part of the task then becomes one of making explicit what is only implicit in Kant (May, 1972, p. 80).

This principle must be applied carefully and within the limitations outlined in the discussion on 'contextual interpretation'. At certain points in the analysis of Harvey's writings, a statement will be taken, and what it implies, as opposed to what it explicitly states, will be explored. The 'exploration of implications', as it may be termed, is more a philosophical than an historical analytical principle. Its aim is to probe the philosophical implications of a statement and so it tends to be an evaluative more than a descriptive device.

An additional principle will prove useful in the interpretation of Harvey's writings. It may be called the principle of 'scholarly inter-action'. Scholarship and research are communal activities conducted in interaction with one's colleagues and with a long tradition of previous writers whose works line the shelves of studies and libraries. At various times, Harvey was arguing against some writers and schools of thought, whereas, at other times, he was developing the views of others. As well, a number of debates dominated the intellectual milieu of Anglo-American geography over the period in which Harvey was writing, for instance, the debate over socially relevant research in the early 1970s. Where necessary, a number of writers, viewpoints and debates will be identi-fied and taken into account in attempting to understand Harvey's views.

Livingstone, in an article entitled 'Some Methodological Problems in the History of Geographical Thought' (1979), noted two general con-siderations which must be acknowledged in any study in the history of geography. First, historians are always forced to select those facts deemed to be significant in the light of their questions. Facts never simply 'speak for themselves'. Secondly, interpretation of facts is neces-sary if the development of geographical thought is to be understood (Livingstone, 1979, pp. 226-7). Indeed, to go one step further than Livingstone, all 'facts' are interpretations in that they are abstracted from any situation by investigators to answer certain questions that they have in mind. Thus, in the examination of Harvey's writings, for instance, philosophical and methodological aspects are emphasised, to the neglect of other aspects, against the background of a particular view of their intellectual milieu.

Livingstone went on to outline a number of specific methodological issues, most of them relating to the study of the history of geography as a discipline rather than to the study of one particular geographer. There are, however, perhaps three points made by Livingstone that ought to be taken into account in a study of Harvey's writings. The first one concerns the danger of overemphasising an aspect of a scholar's work because of the historian's own interests or beliefs (Livingstone, 1979, p. 227). Here, one can only provide one's judgement as best one can and expose it to the critical judgement of other historians. The second point concerns the difficulty of distinguishing the extent to which a scholar reflects contemporary concepts or directs the thought of the time through innovative work (Livingstone, 1979, p. 229). Owing to the difficulty of unravelling the complex interactions between Harvey and geographical thought at the time and to the emphasis in this study on his writings alone, it will be possible only to indicate the points at which Harvey was clearly innovative. The third point of relevance to this study is the need to provide some evaluation of a scholar's writings with respect to 'the correspondence between what he claimed to be doing and what he actually achieved' (Livingstone, 1979, p. 230). At places in the following chapters, it will be possible to break off the *explication du texte* in order to stand back and present such an evaluation.

In Chapters 2, 3 and 4, a chronologically arranged group of Harvey's publications is considered. Each chapter consists of three parts. In the first part, the intellectual milieu of geography at the time is examined, emphasising philosophical developments, and a chronological exposition of Harvey's works is presented in the second part. In part two of Chapter 2, the development of what will later be called a logical empiricist view in Harvey's early works is analysed; in the second part of Chapter 3. Harvey's transition to a Marxist approach and his view of Marx's method is dealt with; and in part two of Chapter 4, Harvey's application of Marx's method to urban geography is examined. The particular philosophical and methodological approaches adopted by Harvey are examined in the third part of each chapter and such issues as his view of theory, of method and of the nature of geography are considered. The conclusions of the study in relation to the questions posed in this introduction are drawn together in Chapter 5.

## Notes

1. This study has been undertaken from the perspective of someone working in New Zealand who has not had first-hand access to David Harvey himself. As it is not a biography, the book includes only a small number of tentative biographical details, which have been built up from hints in the literature.

2. Unless otherwise stated, any emphasis in a quotation is that of the original author.

# 2 LOGICAL EMPIRICIST GEOGRAPHY: SCIENTIFIC EXPLANATION AND THEORY

In 1969, *Explanation in Geography* was published. In this, his first book, which was widely acclaimed, Harvey explored the implications for geography of the logical empiricist philosophy of science. Central to the logical empiricist approach was the deductive-nomological model of scientific explanation and the hypothetico-deductive view of scientific theory. Before the publication of *Explanation in Geography*, Harvey had written a number of articles in which he examined and evaluated a range of models, theories and statistical techniques in geography. Over this period, he had been at the Department of Geography at the University of Bristol. There, he would have engaged in lively debate with such geographers as Haggett, Garner and Chisholm, soon thereafter to be among the leading human geographers in Anglo-America. Harvey had, moreover, participated in conferences and seminars in Britain, Sweden and the United States during the 1960s. *Explanation in Geography* was the culmination of such research and discussion. This chapter examines Harvey's writings in the 1960s against the background of the growing importance of the spatial science approach in Anglo-American geography.

## The Spatial Science Geography of the 1960s

A regional, or 'areal differentiation', approach had dominated Anglo-American geography over the 1930s, 1940s and 1950s.[1] The principal aim of geographical scholarship during this period was seen to be the construction of a synthesis of the characteristics of a region. In Hartshorne's words, 'geography is concerned to provide accurate, orderly and rational description and interpretation of the variable character of the earth surface' (Hartshorne, 1959, p. 21). Systematic studies in human and physical geography were generally seen to be only contributions to a regional synthesis.[2] Each region was assumed to be unique and geography's method was characterised as 'idiographic', that is, 'the intensive study of an individual case' (Hartshorne, 1939, p. 379, 1959, p. 149, note 2).

During the 1950s, a 'spatial science' approach emerged, initially in the United States and notably at the University of Washington, Seattle (Ullman, 1953, 1956; Garrison, 1956, 1959a, 1959b, 1960; Berry and Garrison, 1958a, 1958b). This approach implied that the spatial dimension of the earth's surface was the subject-matter of geography, and the use of statistics and the development of theory was emphasised, particularly, in this case, central place theory. Contemporaneously, also in the United States, Isard pointed out that economics under the influence of Alfred Marshall had largely ignored the spatial component of economic systems (Isard, 1956a, p. 24). Isard founded the Regional Science Association in 1954, seeking to foster interdisciplinary research which gave due weight to the spatial dimension in the development of economic theory and the construction of precise mathematical models (Isard, 1956b, 1960). In that many geographers showed interest in Isard's approach and many of the early papers in regional science were presented at geographical conferences, regional science provided an important stimulus to the development of spatial science geography (Berry and Pred, 1961; Olsson, 1965). Also in the 1950s, Schaefer presented perhaps the earliest critique from the spatial science and logical positivist viewpoint of Hartshorne's account of the nature of geography (Schaefer, 1953).

Spatial science geography came into prominence in Anglo-America in the 1960s and initially appeared as what many called the 'quantitative revolution' (Burton, 1963; Curry, 1967a). Its approach relied upon the scientific method (Bunge, 1962) and encouraged the development of scientific geographic laws (Lukermann, 1961; Lewis, 1965; Golledge and Amedeo, 1968), model-building (Chorley and Haggett, 1967; Cole and King, 1968) and theory construction (Bunge, 1962; Burton, 1963). Spatial science geography was aligned with general systems theory and systems analysis (Chorley, 1962; Berry, 1964b; Chisholm, 1967), often entailing the use of statistical and mathematical techniques (Duncan, Cuzzort and Duncan, 1961; Gregory, 1963; Haggett, 1965; Cole and King, 1968), eventually with the assistance of computer technology (Kao, 1963; Hagerstrand, 1967; Marble, 1967).

In Britain, spatial science geography was dominated by a locational analysis school (Haggett, 1965) seeking to provide a model-based paradigm for geography (Haggett and Chorley, 1967). This school was centred on Cambridge and Bristol. James referred to Haggett, Harvey and Chorley as 'of chief importance as leaders of this new movement' (James, 1972, p. 279). Chorley was a geomorphologist at Cambridge where Haggett initially taught until he took up the second chair of geography at Bristol in 1966. Harvey completed his doctorate at Cambridge and then moved to Bristol in 1961.

In the late 1960s, a group of geographers at Bristol, including Harvey, began to question the validity of certain statistical techniques then in widespread use in geography (Johnston, 1979, pp. 94-5). They pointed out the problems associated with spatial auto-correlation, where the spatial independence of observations required by a technique or model was not able to be met in its application. It was suggested, for instance, that auto-correlation led to biased regression coefficients (Cliff and Ord, 1973).

Earlier, over the late 1950s and early 1960s, G.F. White and his colleagues at the University of Chicago, in attempting to build models of people's perception of and response to environmental hazards, found they had to abandon many of the neo-classical economic assumptions implicit in most contemporary geographic model-building (White, 1945; Roder, 1961; Kates, 1962). Examples of these assumptions included rational decision-making and the availability of perfect information to a decision-maker. In relation to the study of decision-making in a spatial context, Wolpert (1964, 1965) suggested replacing the assumption of rational economic man with that of man as 'satisficer', a concept developed by Simon (1957). These were the beginnings of behavioural geography, which developed out of criticisms of some of the assumptions of spatial science geography but which nevertheless remained closely related to it. The initial period of research in behavioural geography centred on resource-management decisions and environmental perception but it was followed by studies of attitudes and motivations, particularly in the fields of migration, innovation diffusion and political behaviour, especially voting. A number of geographers explored the concept of the 'mental map' as guiding the deliberations of decision-makers (Gould, 1966; Gould and White, 1974), and Cox and Golledge (1969b), in a more general context, suggested that geographers build a corpus of geographical theory around postulates concerning the spatial dimensions of human social and psychological behaviour. This, then, is the background against which Harvey's work over this period must be set.

## David Harvey's Geography, 1961-1969

Between 1961 and 1969, Harvey completed his doctoral thesis and had nine papers and his first book published. His initial research was essentially in historical geography but his attention soon shifted to methodological issues, especially those associated with the use of models and the behavioural approach in geography. In two articles published in 1967 and in *Explanation in Geography* (1969a), Harvey also gave extensive consideration to the formal standards of scientific research.

### Research on Agriculture in Kent

Harvey's doctoral thesis, 'Aspects of Agricultural and Rural Change in Kent, 1800-1900' (1961), was an historical regional study with an economic and agricultural focus.[3] In Part 1, he analysed the regional pattern of change in the hop and fruit industries during the nineteenth century and identified the factors involved in it, for example, fluctuations in demand, market location and the processes of agglomeration and diminishing returns (Harvey, 1961, pp. 166-78). In Part 2, Harvey moved on to examine how

> regional developments in the hop and fruit industries, themselves the result of complex economic and social processes, caused, in turn, significant regional changes in the economic and social structure of rural Kent (Harvey, 1961, p. 251).

Furthermore, he explored how 'each one of the "effects" of land use change could be considered as one of the causes of further land use change' (Harvey, 1961, p. 252). Harvey adopted a causal analysis based upon Max Weber's view (Salaman, 1934; Parsons, 1949) that the scientist uses 'ideal' concepts and sequences to impose order upon an infinitely complex concrete reality. Thus 'a causal argument is a weapon of logic which is essential to understanding' (Harvey, 1961, p. vi).

It is worth noting that Harvey made reference to Marx in this, his initial work, some eleven years before he was to make a serious study of Marx's philosophical and economic thought. Harvey noted that Marx's system of thought viewed the world as 'determined' (Harvey, 1961, p. iv). Then, a few pages later, he wrote that

> the whole of the second part of the thesis . . . is oriented around a Marxist form of analysis whereby the form of the land use is treated as the main determinant of social and economic conditions in rural Kent . . . By 'closing the system' in this way, and isolating the phenomena to be considered, it is possible to gain some insight into the very complex interrelationships which bind one pattern of distribution to another (Harvey, 1961, pp. vi-vii).

By 'Marxist *form* of analysis' (my emphasis), Harvey meant 'closing the system' and viewing one factor as the determining one.

In his concluding remarks, Harvey stated that he had no intention of erecting an 'elaborate mansion of theory' out of the results of his research 'only to find that we are still obliged "to live in the shack outside"' (Harvey, 1961, p. ix). To this extent, he was resisting one of the main thrusts of spatial science geography which, admittedly, was still very much in its infancy. However, Harvey was not unaware of theoretical issues and showed some acquaintance with location theory, making reference to the work of Weber (1928), Losch (1954) and Isard (1956a).

In the introduction to the thesis, Harvey pointed out that Hartshorne (1959) believed that historical geographers should be concerned with the character of areas rather than with the processes giving rise to that character. 'But', Harvey pointed out, 'it is just in the analysis of process that geographical writing on change of location appears most deficient' (Harvey, 1961, p. ii). His thesis attempted to identify and elucidate the processes behind locational change, and therefore, he noted, 'from Hartshorne's standpoint', this took the investigation out of the sphere of geography (Harvey, 1961, p. ii).

Harvey used the gravity model and several simple regression and correlation analyses to explore a number of statistical relationships, for example that between 'density of hops and fruit and rent per acre . . . *c.* 1840' (Harvey, 1961, facing p. 199). Nine years later, Harvey admitted that he misused these techniques, ignoring the problem of spatial auto-correlation and a varying areal scale, but that the results at least portrayed the direction of the relationships. Thus they were merely 'embellishments' to his thesis, and not central to it (Harvey, 1970c, p. 264).[4] Harvey's use of statistical techniques was 'entirely due to the influence of R.J. Chorley' (Harvey, 1961, Acknowledgements), while Allen's (1949) *Statistics for Economists* and Isard's (1956a) *Location and Space Economy* were his referenced sources.

Harvey's doctoral thesis falls between the older regional geography and spatial science geography, with Harvey unhappy with the restrictions of the former but not yet ready to enter fully into the spirit of the latter.

In his first article, 'Locational Change in the Kentish Hop Industry and the Analysis of Land Use Patterns', published in the *Transactions and Papers of the Institute of British Geographers*, Harvey used some of the results of his doctoral research to examine how one land-use pattern developed into another (Harvey, 1963). He suggested that the processes of agglomeration, cumulative change and diminishing returns played an important role and he used statistical analyses similar to those in his thesis. In his conclusions to the article, Harvey made some general theoretical comments on the relevance of his findings for forms of agriculture other than hop cultivation but, as in his thesis, did not at this stage attempt to erect any theory of land-use change.

## Models, Theories and Techniques

In 1965, it became apparent that Harvey was beginning to give serious consideration to the use of models in geography. It was in Sweden during that year that he presented a paper in which he explored the potential use of Monte Carlo simulation models. In the abstract to this paper (Harvey, 1965), Harvey advocated the development of models of general applicability, which could be used to predict future situations. He added a note of caution, however, in pointing out that a great deal more had to be known about such techniques before they could be utilised profitably within geography.

Harvey published three related papers on models in 1966 and 1967. The first was a review article on the use of models in agricultural geography entitled 'Theoretical Concepts and the Analysis of Agricultural Land-use Patterns' (Harvey, 1966a). Published in the *Annals of the Association of American Geographers*, it was Harvey's first publication in the United States. Once again, he acknowledged Chorley's influence, this time upon his view of models (Harvey, 1966a, p. 362, note 3). Harvey surveyed both the economic (Losch, 1954; Isard, 1956a; Garrison, 1959a) and behavioural approaches (Hagerstrand, 1953; Gould, 1963; Wolpert, 1964) and concluded that if model-builders could bridge the gap between spatial models and behavioural models, as well as that between models dealing with time and models dealing with space, then

a general theory of agricultural location which is both operational *and* intuitively satisfying seems a distinct possibility. Such a general theory would be of tremendous utility to geographers, both as a stimulating background for idiographic studies, and as a central pivot in the search for a general theory of spatial interaction. But clearly we are just at the beginning of this quest (Harvey, 1966a, p. 374).

The second paper on models, 'Geographical Processes and the Analysis of Point Patterns: Testing Models of Diffusion by Quadrat Sampling', was published in the *Transactions of the Institute of British Geographers* (Harvey, 1966b). The quadrat sampling technique places a number of quadrats, small areal units, over a map of the region under study. From

this is derived a frequency distribution of the number of quadrats with
0, 1, 2, 3, . . . points contained within them, which may then be compared with a number of theoretically derived distributions. Harvey
tested the ability of five models of spatial diffusion to explain map
patterns from Hagerstand's (1953) study of innovation diffusion among
a Central Swedish farming community. The models tested were taken
mainly from the plant ecology and biology literature, for example,
Douglas (1955), Morisita (1959) and Greig-Smith (1964), although,
in geography, Dacey (1962, 1964) had previously examined similar
models. Harvey noted that it was essential initially to set up a model
which seemed 'intuitively reasonable' because

> we cannot identify an appropriate model simply by fitting probability
> distributions until we find one that has a good fit to the data . . .
> Several different models may give rise to the same probability distri-
> bution (Harvey, 1966b, p. 83).

Since each model represented a different type of process, only one
ought to have been relevant to a particular diffusion situation. Although
Harvey further noted problems in choosing an appropriate quadrat size,
he concluded that the technique was valid and that the models tested
could potentially be used in the analysis of a range of location processes,
not just the diffusion process.

The third paper on models (Harvey, 1967a) was a comprehensive
survey of 'Models of the Evolution of Spatial Patterns in Human Geography' and it formed Chapter 14 of Chorley and Haggett's (1967)
*Models in Geography*. This widely influential book brought together a
number of papers written mainly by young British geographers who in
the near future were to become leading scholars in spatial science geography. Harvey's contribution, together with his two previous articles,
established him as a geographer of growing international reputation.

In the introductory section of his chapter in *Models in Geography*,
Harvey expressed considerable reservations about the highly generalised
theories of spatial evolution of such earlier geographers as Mackinder
(1917) Whittlesey (1929), Dickinson (1951) and Sauer (1952). He
suggested that the informal presentation of these theories did not allow
them to be adequately tested. Harvey went on to present two alternative
approaches to the student of history and geography.

> He can either bury his head, ostrich-like, in the sand grains of an
> idiographic human history, conducted over unique geographic space,

scowl upon broad generalization, and produce a masterly descriptive thesis on what happened when, where. Or he can become a scientist and attempt, by the normal procedures of scientific investigation, to verify, reject, or modify, the stimulating and exciting ideas which his predecessors presented him with. Many historians and historical geographers took the view that 'interpretation depends on scholarship', forgetting that 'without interpretation there can be no scholarship' (Barraclough, 1957). Historical scholarship cannot be conducted in an interpretative vacuum even though many appear to attempt the feat. The first stage of scientific investigation involves the careful testing of theories and generalization, against the facts as they are determined by careful historical investigation (Harvey, 1967a, p. 551).

To Harvey, the role of models in scientific investigation was to formalise a theory, using the tools of logic, set theory and mathematics, and to set out a theory's assumptions and hypotheses in a logical framework so as to eliminate any possible inconsistencies (Harvey, 1967a, p. 552). Harvey concluded that an understanding of the principles and potential of model-building was central to a *'Renaissance* in geographic research' (Harvey, 1967a, p. 597).

The renaissance that Harvey referred to was that of spatial science geography, incorporating both the economic and behavioural approaches. With the publication of these three papers, Harvey entered the geographical methodological debate of his time on the side of the spatial science perspective. He had finally broken with the idiographic approach characterising the older regional geography, effectively labelling it as 'unscientific' (Harvey, 1967a, p. 551). Theory construction, model-building and statistical analysis were central to his view of geography although he recognised the need for careful empirical investigation as its basis. Harvey was not unaware of a number of limitations to and potential misuses of certain modelling and statistical techniques but he called for continuing consideration and application of them.

In two articles published in 1967, Harvey presented the views of the nature of science of 'philosophers and logicians' (Harvey, 1967b, p. 4, 1967c, p. 211) and 'modern analytic philosophy' (Harvey, 1967c, p. 211). Prominent among these views were those of Carnap (1956), Braithwaite (1960), Nagel (1961) and Hempel (1965). Both articles drew upon the manuscript of *Explanation in Geography* (Harvey, 1967c, p. 211, note). The first article, entitled 'Behavioural Postulates and the Construction of Theory in Human Geography', was originally published as a seminar paper by the Department of Geography, University of

Bristol, being published in *Geographia Polonica* in 1970. In it, Harvey
discussed

> the structure and form of scientific theory, . . . the function of such
> theories in geographic explanation, and . . . the implications of this
> mode of analysis for the future development of research in human
> geography (Harvey, 1967b, p. 1).

He began by outlining his version of Hempel's 'covering-law model of
explanation'.

> Hempel (1965) maintains that *all* explanation that presumes to be
> rational can be reduced to the following fundamental form: . . . A
> set of initial conditions, taken in conjunction with a set of law state-
> ments which can be referred to those events, provide a deductive
> explanation for a particular event (Harvey, 1967b, p. 2).

Here, as later in *Explanation in Geography*, Harvey simplified Hempel's
original presentation although he retained its essential characteristics
(Figure 2.1). The function of a scientific theory was 'to provide a coherent and consistent set of law statements which we can use with comparative confidence in such explanations' (Harvey, 1967b, p. 2). There
had been no methodological debate on Hempel's model in geography
up to that time and Harvey assumed, for the purposes of the article,
that the model was relevant to geographic explanation and, quoting
Zetterberg (1965, p. 11), that 'the quest for an explanation is the quest
for a theory'.

Harvey noted that a scientific theory could be regarded as a language
system and referred to Carnap's (1942, 1958) view that such languages
were composed of three features.

> *Pragmatics* describes the conditions under which a language is form-
> ulated; *semantics* describes the relationship between abstract signs
> and symbols contained in the language and empirical phenomena;
> *syntax* describes the logical system of relationships contained within
> the language (Harvey, 1967b, p. 4).

Harvey viewed the syntactical structure of a scientific theory as a
hypothetico-deductive system containing undefined primitive terms
and, logically derived from these, a number of defined terms. Both of
these types of term were to be found in a set of primitive sentences,

Figure 2.1: Harvey's Versions of Hempel's (1965) Covering-law (deductive-nomological) Model of Scientific Explanation (Harvey, 1967b, 2 and 1969a, 36).

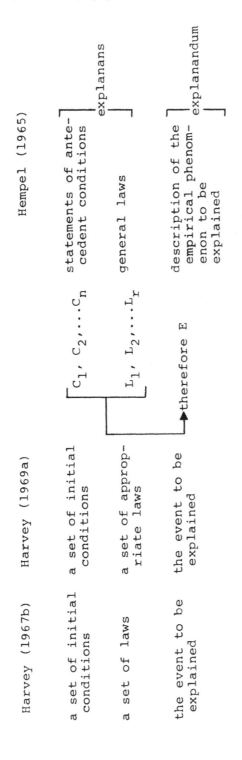

called axioms or basic postulates, and in a set of deductively derived
sentences, called theorems or laws (Harvey, 1967b, pp. 1 and 3-5, 1967c,
p. 212). In Figure 2.2, the relationship between these two types of sent-
ence is presented in diagrammatic form. Such an abstract syntactical
system, or 'calculus', has 'logical coherence but no empirical content'
(Harvey, 1967b, p. 5). A set of designation rules, also called correspondence rules or a text (Harvey, 1967c, p. 213), is needed to relate the
abstract symbols to empirical concepts. As Harvey wrote in the second
article, the 'theoretical language' of the calculus has to be translated
into an 'observation language' (Harvey, 1967c, p. 213). Figure 2.3 sum-
marises Harvey's view of the manner in which an abstract syntactical
system, as presented in Figure 2.2, is given empirical content. A set of
rules to determine the truth condition of, or to verify or falsify, the
sentences contained in the calculus is also necessary (Harvey, 1967b,
p. 5).

Figure 2.2: The Syntax of a Scientific Theory (based upon Harvey,
1967b, 1 and 3-5 and 1967c, 213-14).

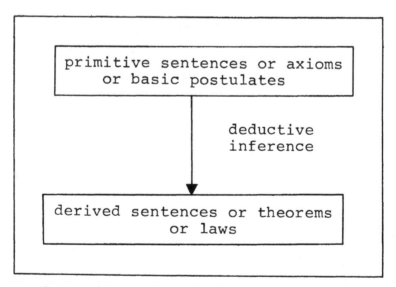

an abstract syntactical system or calculus
     or hypothetico-deductive system

Figure 2.3: The Semantics of a Scientific Theory (based upon Harvey, 1967b, 5 and 1967c, 213).

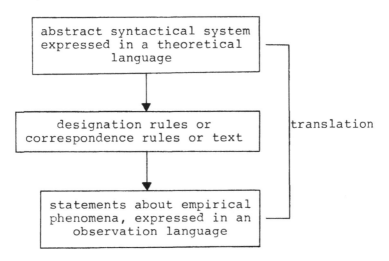

After noting that scientific theory had been only 'weakly' developed in geography, Harvey (1967b, pp. 5ff) went on to discuss what he thought the basic postulates of geographical theory ought to be. He began by pointing out that 'metaphysical speculations' regarding the nature of geography formed the 'mainsprings', that is, the 'directives' or 'regulative principles', of geographical research. Harvey then outlined a 'very general and firmly traditional view of geography', that of 'chorology; . . . knowledge of the varying expression of the different parts of the earth's surface' (Harvey, 1967b, p. 6). Given this view of geography, Harvey believed that there were two distinct types of theory in geography. The first type consisted of theories developed by other empirical sciences relating to temporal processes whereas the second consisted of theories indigenous to geography relating to spatial form.

The fact that postulates indigenous to geography were those concerned with 'location, nearness, distance, pattern, morphology and so on' (Harvey, 1967b, p. 7) emphasised the close relationship between geometry and geography, the former providing abstract calculi for the theory of the latter. The development of laws describing spatial form, areal co-variance and so on, was the aim of indigenous theory. But 'laws of spatial pattern do not necessarily tell us anything as regards process' (Harvey, 1967b, p. 8), and so human geography must derive theoretical postulates from the social sciences, especially from psychology, economics and sociology. Harvey noted Wolpert's (1964) introduction of the idea of 'satisficing behaviour', derived from the work of Simon (1957, 1966), a psychologist, although Harvey pointed out that both Sauer (1941) and Brunhes (1912) had previously referred to a similar concept (Harvey, 1967b, pp. 9-10). The possibilities of reducing all the theories of the social sciences to a small number of basic postulates and, in turn, reducing these to purely physical postulates (Carnap, 1956; Hempel, 1965, Chapter 7; Freeman, 1966) were discussed by Harvey, along with various objections to such reductions (Harvey, 1967b, pp. 11-12).

Since spatial patterns in human geography could be regarded as being the immediate consequence of human decisions, Harvey went on to examine a 'decision making framework' (Harvey, 1967b, pp. 13-24). He surveyed three approaches which utilised models in the analysis of decision-making processes. The first approach utilised normative models, that is, models describing 'how people ought to behave given certain goals' (Harvey, 1970b, p. 35).[5] Here, Harvey reviewed microeconomic theory (von Thunen, Losch and Weber) and normative decision models, especially game theory (Gould, 1963). Analytic models, describing 'how people ought to act given our knowledge of human behaviour' (Harvey, 1970b, p. 35), were used in the second approach. Some of these psychological models had been applied in geography by, for example, Gould (1966) and Saarinen (1966), the former in investigating mental maps, the latter in examining perception of drought hazard. The third approach adopted the descriptive models of, for instance, the behavioural psychologists, dealing with 'how people in fact behave in response to certain stimuli' (Harvey, 1970b, p. 35). Harvey concluded from his survey that

> at the present time the level of understanding and control which behavioural science as a whole gives us over decision making processes is not sufficient for any very powerful synthesis from behavioural postulates to spatial pattern to be developed. As behavioural science itself progresses so we may expect more powerful behavioural postulates to be generated . . . Geographic theory must rest upon explicit assumptions about human behaviour. The sooner this problem is tackled directly the better (Harvey, 1967b, p. 24).

The second of Harvey's 1967 articles dealing with the nature of science was, significantly, an editorial to a special issue of the *Journal of Regional Science*. Harvey's task was to introduce a number of essays by geographers contributing to some aspect of geographic theory (Harvey,

1967c). The editorial was entitled 'The Problem of Theory Construction in Geography' and in it Harvey chose to discuss the structure of scientific theory as outlined in the previous article and summarised in Figures 2.2 and 2.3. Owing perhaps to the brevity of the introductory statement, Harvey did not deal with Hempel's covering-law model of scientific explanation.

Harvey saw the main theoretical problem of geography to be the translation of 'temporal relationships into their spatial equivalent' (Harvey, 1967c, p. 213). There were two possible ways to achieve this, he believed. Human geographers could attempt a direct derivation of spatial form from, for example, economic or psychological postulates. Alternatively, an attempt could be made to develop general theory appropriate for discussing both spatial and temporal systems. This might be achieved by developing a calculus which could be given both a spatial and a temporal interpretation. Harvey pointed to Curry's (1967b) article as exemplifying this approach (Harvey, 1967c, pp. 213-14). However, Harvey conceded that an adequate text or interpretation had not yet been provided for economic theory. In fact, general theory and reliable verification procedures were absent in all the social sciences. Harvey concluded that, in general, geography's theoretical constructs were 'a chaos of vague postulates, fuzzy argument, and often unwarranted conclusions . . . The search for an adequate corpus of geographic theory is not many years old' (Harvey, 1967c, pp. 214-15).

In these two articles, Harvey had shifted the focus of his methodological reflections from the subject of models to that of the nature of scientific explanation and theory. Whereas other geographers were introducing ideas, models and theories into geography from cognate disciplines such as economics and psychology, Harvey, in these articles, was introducing material from the philosophy of science into the methodological considerations of geography. In doing so, his concern was for greater logical control and power in geographical explanation by means of a more formal structuring of the theories used by geographers. He was treating a particular view of the nature of science, that developed by what he called 'modern analytic philosophy', as the standard by which geography's theories should be judged. It was apparent to Harvey that a lot of work was required before these theories would meet such a standard.

In his next two articles, both published in the *Transactions of the Institute of British Geographers* in 1968, the focus of Harvey's work returned to the use of statistical models in geography. Originally presented to the Statistics Study Group of the Institute of British Geographers in

January 1967 (Harvey, 1968a, p. 94, note 1), the first of these publications was entitled 'Some Methodological Problems in the use of the
Neyman Type A and the Negative Binomial Probability Distributions
for the Analysis of Spatial Point Patterns' (Harvey, 1968a). The general
problem that Harvey was concerned with was that of 'applying *a priori*
analytical knowledge such as that contained in pure mathematical systems, to *a posteriori* synthetic empirical problems' (Harvey, 1968a,
p. 85). The two models examined in the paper were based on probability
theory which 'provides us with a powerful abstract logical system which
we may use as a *model* for describing and analysing geographical distributions' (Harvey, 1968a, p. 85). These were two of the more successful
models tested in an earlier article (Harvey, 1966b) and were again examined in relation to the analysis of diffusion processes.

Harvey noted how the assumptions of the models severely circumscribed the choice of the quadrat size to be used in sampling a study
area. Furthermore, 'the circumstances under which we can validly apply
either of these probability distributions appear to be very limited, since
the criteria for the application of these mathematical models are very
strict' (Harvey, 1968a, p. 89). However, research in plant ecology and
geography (Rogers, 1965; Harvey, 1966b) had demonstrated that both
theoretical distributions gave successful fits in a wide range of circumstances. Harvey went on to list three interrelated reasons which he
believed could explain this.

(1) Although the criteria for the application of the mathematical
model are not fully met in the real world, they are sufficiently well
approximated for the model to give a good fit. (2) The same mathematical model may be derived by applying radically different criteria.
(3) The sampling system (particularly with respect to quadrat size)
is producing spurious results (Harvey, 1968a, p. 89).

Harvey went on to present an example where, given random variation
in the susceptibility of a population to accepting information, using a
small quadrat size for sampling would yield a frequency distribution
similar to a simple Poisson (random) model whereas an increased quadrat size would yield a negative binomial type of distribution (Harvey,
1968a, pp. 90-1). Harvey also pointed out that Neyman and Scott (1957)
had attributed some of the good fits obtained with the negative binomial model to spatial auto-correlation, whereas Skellam (1958) had
noted how the Neyman type A distribution could be derived from the

misapplication of the quadrat sampling technique (Harvey, 1968a, pp. 91-2). Thus a good fit with a particular model did not necessarily mean that the process which the model was originally constructed to simulate was in fact the process at work in the situation under study. Harvey suggested that one way of discriminating among alternative theoretical interpretations of the same model in geography was to develop a more adequate location theory. But 'our ability to use mathematical models is tending to outrun the construction of geographical theory' (Harvey, 1968a, p. 94). Harvey distinguished between 'mere model "consumption"' and 'the construction of powerful theory'. He concluded that theory construction and mathematical model-building must proceed at the same pace if geographers were to achieve any 'deep understanding' of the processes which created spatial patterns (Harvey, 1968a, p. 94).

Problems associated with deducing spatial patterns from a knowledge of temporal processes formed the focus of the second article by Harvey published in 1968, 'Pattern, Process, and the Scale Problem in Geographical Research' (Harvey, 1968b). The major problems examined were those of scale and spatial auto-correlation. Harvey argued that different processes were significant to the understanding of spatial patterns at different scales. However,

> we have no measure of the scale at which a particular process has most to contribute to the formation of a spatial pattern and our notions regarding the scale problem remain intuitively rather than empirically based (Harvey, 1968b, p. 72).

Harvey went on to suggest that the scale problem could be used creatively.

> An accurate knowledge of the scale at which a process is most effective in determining spatial form may be used to determine the scale for measuring that form. The analysis of the properties of spatial pattern at a variety of scales can also help to identify the scale at which a particular process is most effective. It is thus suggested that an iterative analysis going from process to pattern back to process, and so on, may provide useful insight into the scale problem (Harvey, 1968b, p. 72).

With 'perfect information' on the scale at which a process operated and with 'specific knowledge' about the relevant distance function, Harvey believed it would eventually be possible to predict the resultant spatial pattern of events or objects governed by such a process, and to specify the scale at which the real world pattern should be measured. 'But', he was forced to admit, 'as yet we do not possess this knowledge' (Harvey, 1968b, p. 73). However, Harvey went on to suggest that the identification of the scale at which a pattern deviated most from randomness might provide insight into the structure of the pattern and into the manner in which scale and process were interrelated (Harvey, 1968b, p. 74).

Harvey also presented a method of estimating the extent of auto-correlation in spatial patterns and he noted the potential use of spectral analysis in analysing spatial patterns (Harvey, 1968b, pp. 75-7). He concluded the article with the following prediction:

> It may well be that, in the near future, spatial auto-correlation and the scale problem . . . will be regarded as the key problems whose joint solution paved the way for more penetrating analysis of the relationship between temporal process and spatial form (Harvey, 1968b, p. 78).

In the two articles published in 1968, Harvey demonstrated that there was a range of difficulties associated with the application of certain statistical techniques in geography. At this stage, he appeared confident that geographers could overcome most of these difficulties, although not without sustained and creative effort. In the second article, Harvey himself suggested an innovative approach to two 'problems' which could potentially aid geographers in their task of analysing the relationship between form and process.

## Explanation in Geography

Harvey's major publication in this period was *Explanation in Geography*. In it, he brought together his previous research and, extending his purview considerably, sought to construct a coherent methodological framework. He noted in the preface to the book that he had written it mainly to educate himself after he had concluded that more than techniques were involved in the 'quantitative revolution' of the 1960s. It was a 'philosophical revolution' that was taking place and Harvey decided he had to abandon certain assumptions he had accumulated in his six years of 'indoctrination of what I can only call "traditional" geography at Cambridge' (Harvey, 1969a, p. vi). These assumptions were based on the view that things were unique and that human behaviour could not be measured.

Harvey stated that he believed that there was nothing wrong with the aims and objectives of 'traditional' geography but that geographers needed to 'take advantage of the fantastic power of the scientific method', for it was the 'philosophy of the scientific method' which was implicit in quantification (Harvey, 1969a, pp. vi-vii). Harvey thus decided to concentrate upon 'the standards and norms of logical argument and inference which geographers ought to accept in the course of research'. In short, he had become interested in 'the role of scientific method . . . in geography'. The issue of quantification, as such, faded into the background (Harvey, 1969a, p. vii). Harvey also believed that mathematics and statistics, the 'sharpest tools' of science, could easily be misapplied and misunderstood. In order to control the use of these tools, 'we must understand the philosophical and methodological assumptions upon which their use necessarily rests' (Harvey, 1969a, p. viii).

Harvey wrote the preface to *Explanation in Geography* in March 1969, but had completed the manuscript for the book in June 1968 (Harvey, 1969a, p. viii). He noted in the preface that he had changed his opinion on several issues over the intervening period and, because of this, considered the book to be merely

> an interim report — one person's view at a particular point in time.
> I do not wish it to form the basis for some new kind of orthodoxy
> . . . My aim is to open up the field of play rather than to close it off
> to future development (Harvey, 1969a, p. viii).

Within the main body of the text of *Explanation in Geography*, Harvey presented the book as part of a search by geographers for a new paradigm for their discipline (Harvey, 1969a, p. 18). Harvey's contribution to this search entailed, first, the presentation of the view of 'the logical positivists and philosophers of science' (Harvey, 1969a, p. 6) on the nature of scientific explanation, theories and models, and, secondly, the development of a 'strategy' for the construction of geographic theory through the use of models (Harvey, 1969a, p. 176). *Explanation in Geography* may, on the basis of these two aspects of Harvey's contribution, be viewed as consisting of two major sections. The first, eleven chapters long (Chapters 2 to 12), deals with topics such as scientific explanation, theories, hypotheses, laws and models. The second major section, also eleven chapters long (Chapters 13 to 23), deals with artificial model languages and the use of models for description and explanation in geography. There is also an introductory chapter on philosophy and methodology in geography and a concluding chapter in which Harvey comments on how methodology could be used 'to promote a more concise philosophy of geography' (Harvey, 1969a, p. 481).

*Explanation in Geography: Scientific Explanation and Theory.* Harvey's method of presentation throughout *Explanation in Geography* often follows a pattern. Having started with a brief introduction to a topic, such as scientific explanation, he would go on to outline the views of 'the logical positivists and the philosophers of science'. Geographical research would then be evaluated in the light of these views and Harvey would conclude by looking at the ways in which geography could be developed along the lines indicated by the philosophers of science.

In order both to outline the main elements of *Explanation in Geography* and to examine its underlying philosophical basis, the following analysis of the book will give attention to these issues: Hempel's covering-law model of scientific explanation, the hypothetico-deductive system of scientific theory and the nature of models; the evaluation of explanation, theories and models in geography and a strategy for the development of scientific theory in geography; the relationship between methodology and philosophy in geography; and, finally, the nature of geography.

Harvey outlined three ways of constructing a scientific explanation. The 'deductive-predictive' approach, based on explanation in physics, had been developed by Braithwaite (1960), Nagel (1961) and Hempel (1965). The 'relational' approach, which related the event to be explained (for example, the behaviour of planets) to other events (for example, apples falling from trees), had been developed by Toulmin (1960) and Hanson (1965). 'Explanation by way of analogy' was the approach of Workman (1964) (Harvey, 1969a, pp. 13-15). Harvey indicated that he intended to make greater use of the 'deductive-predictive' mode 'mainly because of its simplicity' (Harvey, 1969a, p. 15). Later in the book, Harvey considered 'verstehen' as a mode of validating arguments. He noted that Weber (1949) and Winch (1958) had proposed that explanation in the social sciences involved 'understanding a particular event by empathy with the . . . individuals involved in the event', by 'putting oneself in another person's shoes', an operation known as 'verstehen' (Harvey, 1969a, p. 56). Harvey believed that although 'verstehen' may be fundamental to the imaginative creation of hypotheses,

it does not add to our store of knowledge because it amounts to applying knowledge already validated by personal experience. Nor "does it serve as a means of verification. The probability of a connection can be ascertained only by means of objective, experimental, and statistical tests" [Abel, 1948] (Harvey, 1969a, p. 59).

Harvey further damned 'verstehen' by associating it with Hartshorne's regional approach (Harvey, 1969a, p. 71).

Although he went on to examine carefully and eventually to recommend the 'deductive-predictive' approach to explanation, Harvey was not uncritical of certain fundamental aspects of it. He pointed out that it tended to ignore explanation as an activity, as a process and as an organised attempt at communicable understanding. It was a 'formal' approach, unlike Kuhn's (1962) 'behavioural' approach. As well, analytic philosophy treated experience as 'some set of statements about reality'. It did not give an account of the interpretation to be given to experience (Harvey, 1969a, p. 9). With these observations providing a caveat to his presentation, Harvey went on to consider Hempel's model of scientific explanation, which is central to the 'deductive-predictive' approach (Figure 2.1).

Harvey noted that Hempel himself called his model of explanation a 'deductive-nomological' form of explanation (Harvey, 1969a, p. 36). The term 'deductive-nomological' refers to the two most important aspects of the model. First of all, the type of logical inference used in the model is that of deduction. That is, given a set of initial conditions and a set of appropriate laws, it is possible to deduce either that a particular event has happened or that it will happen. Thus, 'in this form of explanation, prediction and explanation are symmetrical' (Harvey, 1969a, p. 37). Although deduction cannot add anything to our knowledge that is not already contained in the premises, it 'ensures the logical certainty of the conclusion' if both premises are true (Harvey, 1969a, p. 37). With inductive inference, true premises do not ensure a true conclusion.

Hume made the point that because we conduct an experiment a thousand times and get the same result, we cannot infer with certainty that the next experiment conducted under the same conditions will necessarily yield the same result . . . There is no logical justification for extending belief in the premises to belief in the conclusions (Harvey, 1969a, p. 37).

The second important aspect of Hempel's model referred to by the term 'deductive-nomological' is its use of laws. A law or set of laws must form one of the premisses of the deductive inference, whether the model is being used to predict or to explain an event. Harvey pointed out that 'logicians and philosophers' have suggested that two criteria be used to identify a law. First of all, a law is a universal statement 'unrestricted in its application over space and time' (Harvey, 1969a, p. 101). Such a rigid interpretation of the character of laws 'would probably mean that scientific laws would be non-existent in all of the sciences' so this criterion is usually relaxed. Harvey noted Ackoff's (1962, p. 1) view that 'the less general a statement the more *fact-like* it is; the more general a statement, the more *law-like* it is. Hence, facts and laws represent ranges along the scale of generality' (Harvey, 1969a, p. 104). Secondly, a law is a statement related to a surrounding structure of theory. The theory defines the terms used in the law and ensures that the law statements used in explanation are consistent with respect to each other (Harvey, 1969a, p. 90). As far as its verification is concerned, a scientific law has to be established as true by reference to 'empirical subject-matter' as well as be supported by other established laws (Harvey, 1969a, p. 105).

Harvey noted that Braithwaite (1960) characterised the systematic organisation of scientific knowledge as a 'hypothetico-deductive system' in which a set of hypotheses formed a hierarchically ordered deductive system, higher level hypotheses serving as premisses (Harvey, 1969a, p. 36). As in his earlier articles (Harvey, 1967b, 1967c), Harvey referred to the higher level hypotheses as postulates or axioms, and to the lower level hypotheses as laws or theorems (Harvey, 1969a, p. 390). This mode of organising scientific knowledge is summarised in Figure 2.2.

Harvey outlined two alternative routes to the construction and verification of scientific laws and theories. The first, the 'Baconian' route, was inductively based and assumed that the processes of ordering and structuring data were independent of theory. The second route, which Harvey saw as being based largely upon deductive inference, recognised the *a priori* role of theory with regard to data (Harvey, 1969a, pp. 32-6). Harvey's view, as put forward in *Explanation in Geography*, of the origin, structure and verification of scientific laws and theories, and of the relationship between theories, laws and the deductive-nomological model of scientific explanation, is summarised in Figure 2.4. This diagram of the scientific method is based on what Harvey called the 'second route' to scientific explanation, but it is more complex than the account put forward at that particular point in *Explanation in Geography* and more accurately represents the view presented throughout the first half of the book.

Figure 2.4: The Origin, Structure and Verification of Scientific Laws and Theories (based upon Harvey, 1969a, esp. 3-5, 16-22, 32-40, 87-96, 100-6, 114 and 350).

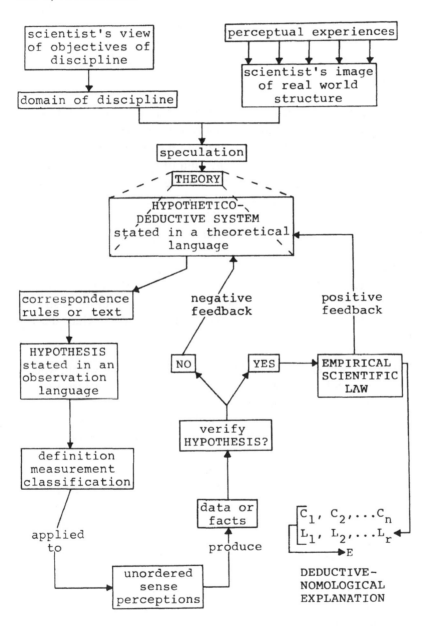

Harvey believed that the origin of a theory was to be found in 'spec-ulation'. He noted that

> Theories are, in Einstein's phrase, "free creations of the human mind". Any speculative fantasy may thus be regarded as a theory of some sort . . . Metaphysical speculation has been a source of stimulating ideas in all areas of scientific research (Harvey, 1969a, p. 87).

Speculations of use in science are speculations about those objects and events viewed as being within the 'domain' of a scientific discipline. Scientists also have an image of the 'real world structure' in their minds which determines to a large extent the kind of speculation that is pos-sible (Harvey, 1969a, pp. 33-4). This 'intuitive picturing' of reality has been built up through a complicated interaction between perceptual experiences, language and thought (Harvey, 1969a, pp. 18-22, 34). Thus, as Figure 2.4 illustrates, a postulated theory assumes both a prior view of the domain of a discipline and the scientist's intuitive picturing of reality.

Harvey believed that a theory should have a logical structure which ensures consistency and a set of statements which connect the abstract notions contained in the theory to 'sense-perception data'. Once this is achieved,

> The theory will enable us to deduce sets of hypotheses which, when given an empirical interpretation, may be tested against sense-perception data. The more hypotheses we can check in this fashion, the more confident we may feel in the validity of the theory pro-vided, of course, that the tests prove positive (Harvey, 1969a, p. 35).

The theory should be formally stated in a theoretical language, for example, mathematics. It is the unambiguity and precision of the theor-etical language which give science much of 'its power of objectivity and universality', although Harvey was aware that, in order to attain uni-versal validity or scope, such a language is, paradoxically, 'restricted in the range of experience to which it can refer' (Harvey, 1969a, p. 22). Furthermore, Harvey noted that 'it is comparatively rare for theories in either the natural or social sciences to be stated in a completely formal manner' and he outlined a classification of theoretical structures in terms of their degree of formalisation (Harvey, 1969a, pp. 96-9).

Through the application of a text or set of correspondence rules, a theorem, or theoretical law, will, upon translation, become a hypothesis containing 'empirically identifiable subject-matter' (Harvey, 1969a, p. 91).[6] If it is eventually verified, the hypothesis will attain the status of an 'empirical law' (see Figure 2.4). This type of scientific hypothesis must be a universal or, at the very least, a general statement so that it can take its place as a theorem in the formalised structure of a theory. When verified, it will help to confirm the reliability of the theory. As Harvey put it, quoting Braithwaite (1960),

> The empirical testing of the deductive system is effected by testing the lowest-level hypotheses in the system. The confirmation or refutation of these is the criterion by which the truth of all the hypotheses in the system is tested (Harvey, 1969a, p. 100).

To confirm or refute a hypothesis, the procedures of definition, classification and measurement are necessary. Harvey called these procedures 'operational filters' (Harvey, 1969a, p. 305). In surveying the various types of definition, he concluded that the most consistent type was 'operational definition' where, quoting Stevens (1935, p. 323), definition 'consists simply in referring any concept . . . to the concrete operations by which knowledge of the thing in question is had' (Harvey, 1969a, p. 303). Measurement is a way of structuring observations and, in relation to quantification, Harvey put forward the view that

> In science quantity is not the polar opposite term to quality – it should better be regarded as a superior form of quality. In principle it ought to be possible, in all spheres of our understanding, to improve the quality of our understanding by some form of quantification. It is true that there are many areas of our understanding which prove intractable to quantification, but in such situations it appears that the intractability comes from our basic lack of understanding, our failure to conceptualise the situation reasonably, rather than from anything inherent in the situation itself (Harvey, 1969a, p. 308).

Classification was defined by Harvey as 'the basic procedure by which we impose some sort of order and coherence upon the vast inflow of information from the real world' (Harvey, 1969a, p. 326).

'Data' or 'facts' are produced when the procedures of definition, measurement and classification are applied to 'unordered sense perceptions' (Harvey, 1969a, p. 350). As illustrated in Figure 2.4, the verification of a scientific hypothesis involves its comparison with these data.

Such empirical verification must therefore rely upon inductive inference, the reliability of which has been discredited. Harvey outlined
three main schools of thought on this problem of verification. Nagel,
Carnap and Hempel supported a probability approach to select the
most confirmed hypothesis. Popper (1968) put forward the idea that
a hypothesis should be assumed to be true and that the scientist should
set out to falsify it. Kuhn's (1962) notion of paradigms assumed that a
hypothesis was verified according to paradigm-determined rules, which
may or may not entail logical justification (Harvey, 1969a, pp. 38-40).
Harvey noted that there were difficulties with each of these views and
concluded that this situation 'gives the lie to those who claim total
objectivity for knowledge accumulated by way of the scientific model'
(Harvey, 1969a, p. 40). Once a scientific hypothesis has been verified
by some method to the satisfaction of scientists, it becomes an empirical law. This has a 'positive feedback' effect upon the original theory,
strengthening the range of support for its validity. Alternatively, as
Figure 2.4 illustrates, proving a scientific hypothesis false has a 'negative
feedback' effect and the theory should be reassessed (Harvey, 1969a,
p. 34).

For Harvey, the development of empirical laws was also important
because of their potential use in Hempel's deductive-nomological model
of scientific explanation. The origin and generation of a scientific law
within the framework of a hypothetico-deductive system, and its verification by comparison with empirical data, was a strategy to minimise
reliance upon inductive inference for the support of an empirical law.
In doing this, Harvey's statement in *Explanation in Geography* went far
beyond that of any previous geographer. Harvey was offering spatial
science geographers a coherent methodological framework within which
they could assess their scientific contribution.

When he turned his attention to the topic of models in *Explanation
in Geography*, Harvey began by surveying their many functions, as
psychological, normative, organisational, explanatory and constructional
devices (Harvey, 1969a, p. 141). He developed a schema to represent
the various functions of models as a contribution towards clarifying two
methodological problems of model use, namely, how to establish, first,
which function a model is performing, and secondly, the appropriate-
ness of a particular model for a particular function (Harvey, 1969a,
pp. 141-3). Harvey went on to survey various definitions of a model, including that of Braithwaite (1960), Brodbeck (1959) and Nagel (1961)
who were of the view that 'a model for a theory T is another theory M
which corresponds to the theory T in respect of deductive structure'

(Harvey, 1969a, p. 145). Harvey then discussed an important distinction, that between *a posteriori* and *a priori* models. An *a posteriori* model is 'developed in order to represent the theory' (Harvey, 1969a, p. 151). It may be used in the process of testing a theory, whether it be in definition, classification, measurement, data collection, data representation or verification. An *a priori* model is a 'ready-made calculus' applied to 'some aspect of the real world' and 'we may, from the structure of the calculus, infer the structure of the theory' (Harvey, 1969a, p. 153). This meant that an *a priori* model could be used to organise and manipulate an initially postulated theory. Both the model and the theory are calculi in that both consist of axioms and theorems in a deductive inferential relationship. In this way, the model corresponds to the theory 'in respect of deductive structure', meeting the criterion of Braithwaite, Brodbeck and Nagel. Such an *a priori* model helps to formalise the initial theory and allows the deduction of testable hypotheses upon the provision of a text or set of correspondence rules (Harvey, 1969a, p. 154).[7]

Throughout the first major section of *Explanation in Geography*, Harvey assessed geography's explanatory procedures and theoretical constructs in the light of Hempel's deductive-nomological model of scientific explanation and the hypothetico-deductive structure of theory. His approach both reflected and directed the then current trends in geographical thought. He noted that, in the past, 'geographers have not argued about explanatory form but have argued about objectives' (Harvey, 1969a, p. 64). 'Most controversy has thus centred on philosophical rather than methodological issues' (Harvey, 1969a, p. 67). Harvey pointed out that Hartshorne's views, widely influential over the previous 30 years, had relied upon Hettner's work who, in turn, was influenced by the 'German school of historiographers'. This meant that

The philosophical underpinnings of Hartshorne's methodology . . . appear to relate to the philosophy of history in the latter half of the nineteenth century rather than to the philosophy of science in the mid-twentieth. It is possible . . . that the uniqueness thesis and the idiographic method can be given a modern defence, but such a defence needs to parry the challenge posed to the uniqueness thesis in modern philosophy of science (Harvey, 1969a, p. 65).

Hartshorne's espousal of the 'idiographic method', defined by Harvey as 'the exploration of particular connections', as opposed to the 'nomothetic method' which is ˙concerned with establishing generalisations'

(Harvey, 1969a, p. 50), led Hartshorne to the view that

> we are essentially concerned with unique collections of events and
> objects . . . Locations, it was argued, are unique . . . The description
> and interpretation of what existed at those unique locations could
> not be accomplished by referring to general laws. It required, rather,
> understanding in the sense of empathy or *verstehen*, i.e. the employ-
> ment of the idiographic method (Harvey, 1969a, pp. 70-1).

Such a view, as well as the desire to prevent a return to the environmental determinism of early twentieth century geography, had made
geographers wary of the formulation of laws. The extreme complexity
of geography's subject-matter also made the formulation of scientific
laws very difficult (Harvey, 1969a, p. 174). In spite of these factors,
however, Harvey believed that the work of Bunge (1962) and Haggett
(1965) indicated a trend towards treating events *'as if'* they were 'subject to scientific law' (Harvey, 1969a, p. 109).

Quoting Ballabon (1957, p. 218), Harvey admitted that geography
was 'short on theory and long on facts'. Despite this, geographers had
developed a range of concepts and principles which could 'ultimately
act as the basic postulates for theory' (Harvey, 1969a, p. 117). These
included derivative concepts from, for example, economics, psychology,
sociology and physics, and indigenous concepts describing and explaining geographical phenomena (Harvey, 1969a, pp. 116-27). As far as
models were concerned, Harvey believed that there was 'a great deal
of variation among geographers regarding the meaning of the term and
the practical application of the model concept in empirical research'
(Harvey, 1969a, p. 162). He viewed the Davisian *a priori* model of the
erosion cycle as only a partial representation of a theory but it had been
treated in geography as a complete theory. Harvey also noted how geographers had given the negative binomial probability distribution model
a wide range of different interpretations. He went on to criticise his
own earlier view that a model was an interpretation or representation
of a theory (Harvey, 1967a, pp. 552-4). Rather, quoting Chorley (1964,
p. 128), Harvey pointed out that 'a model becomes a theory about the
real world only when a segment of the real world has been successfully
mapped into it' (Harvey, 1969a, p. 162).

Harvey generally concluded that

> Explanation in geography has, until recently, remained a process of
> applying intuitive understanding to a large number of individual cases.

Scientific theories have not, by and large, been explicitly developed, laws have consequently not been formulated, and the usual require- ments of scientific explanation have not been met (Harvey, 1969a, p. 173).

He went on to state that

There is no reason in principle why laws should not serve to explain geographical phenomena, or theories of considerable explanatory power be constructed. Explanations which conform to the rules of scientific explanation as generally conceived of can, in principle, be offered. This is our central conclusion (Harvey, 1969a, p. 173).

But, noted Harvey, 'the main difficulty comes with the implementation of this conclusion'. Lack of understanding by geographers and the ex- treme complexity of their subject-matter meant that theories of any great explanatory power would be a long way off. Harvey decided that geography needed to develop a 'weaker' paradigm of explanation and theory, although one 'not entirely unrelated' to the 'scientific' para- digm. Central to such a geographical paradigm should be 'the willingness to regard events *as if* they are subject to explanation by laws' (Harvey, 1969a, p. 174). Taking as his starting point Chorley and Haggett's (1967) argument for a model-based paradigm for geography, Harvey also advocated the use of *a priori* model concepts directed to the con- struction of theory. He believed that this would allow geographers to venture predictions in the absence of complete theory and would indi- cate the appropriate theory for a given situation.

There are vast ready-made calculi which can be given some geographic interpretation with great benefit. But in order to comprehend the results we must be prepared to learn the language of the calculus we are proposing to use, understand its properties, modify it where necessary, and consciously manipulate it to suit our own needs (Harvey, 1969a, pp. 175-6).

These comments brought Harvey to the close of what has been referred to as the first major section of *Explanation in Geography*. In the second major section, he went on to explore the potential use of mathematics, geometry and probability theory in providing model languages for geography. He then outlined how models could be utilised for descrip- tion and explanation in geography. In relation to explanation, Harvey surveyed the use of cause-and-effect models, 'temporal modes' of explanation, functional explanation and systems analysis. When examining functionalism, he pointed out that

> In the twentieth century functionalism has tended to mark a reaction against crude cause-and-effect determinism (causalism) and nineteenth-century positivism. It has sought to replace the language of cause and effect by a language stressing interrelationships, and to provide an alternative mode of explanation to those mechanistic forms characteristic of physics (Harvey, 1969a, p. 439).

In anthropology, Radcliffe-Brown (1952) had utilised a structural-functional approach to analyse the functioning and development of social structures which consisted of 'a *set of relations* amongst *unit entities*' (Harvey, 1969a, p. 440). Radcliffe-Brown, however, saw functionalism as a 'working hypothesis' and did not commit the error of the 'functional fallacy — the view that *everything* in society has a function and that it could therefore be understood only in terms of that function' (Harvey, 1969a, p. 440). Thus, for Harvey, to attack functionalism as a philosophy was not to attack it as a methodology (Harvey, 1969a, p. 445). The emphasis of methodological functionalism on interrelatedness, interaction and feedback within and among complex organisational structures or systems had proved 'extremely rewarding' in geography (Harvey, 1969a, pp. 441-5). But the functionalist method was seen by such philosophers of science as Hempel (1959), Braithwaite (1960) and Nagel (1961) as but an approximation to more efficient forms of explanation. Harvey concluded that, because of the complexity of the world, geographers had to continue to utilise functionalist methodology, but only until 'full-blown theory' had been developed. And, above all, 'the danger we must here avoid at all costs is that mortal inferential sin of erecting an *a priori* functional model into full theory without knowing it and without the necessary confirmatory evidence' (Harvey, 1969a, p. 446).

In the chapter on systems, Harvey argued that the concept of a system provided the key to explanation.

> Any explanation involves the isolation of certain events and conditions and the application of certain law (or law-like) statements to show that the events to be explained must necessarily have occurred, given that certain other events or conditions occurred. Isolating events and conditions in this way amounts to defining a

*closed system* (Harvey, 1969a, p. 447).

Thus Hempel's deductive-nomological model of scientific explanation necessarily assumed a closed system. For Harvey, the structure of a system consisted of a set of elements linked together in one or a number of ways. He believed that the most difficult problem in using systems concepts concerned defining a particular system's boundaries.

> In some cases the boundaries are fairly self-evident . . . In other cases we are forced to *impose* boundaries in some fashion . . . The choice of boundaries, however, may have a considerable impact upon the results obtained from systems analysis . . . Both approaches are reasonable; and which is chosen depends mainly on what is feasible and what we wish to investigate (Harvey, 1969a, p. 457).

There was a long history of 'systems thinking' in geography, according to Harvey. However, it had been bound up with the functional approach, organismic analogies, the concept of regions as complex interrelated wholes and the ecological approach. It was not until the early 1960s that such geographers as Chorley (1962), Ackerman (1963) and Berry (1964a) had directed attention to the need for a reformulation of geographic objectives in formal systems terms. In some instances, for example Chorley (1962) and Berry (1964b), systems concepts had been used to develop new theoretical formulations (Harvey, 1969a, pp. 467-8). But Harvey believed that the employment of systems concepts and systems analysis had not yet achieved 'powerful operational status in geography' partly because of the complexity of the method and its use of mathematical techniques beyond the reach of most geographers. He concluded,

> Whatever our philosophical views may be, it has been shown that methodologically the concept of the system is absolutely vital to the development of a satisfactory explanation. If we abandon the concept of the system we abandon one of the most powerful devices yet invented for deriving satisfactory answers to questions that we pose regarding the complex world that surrounds us. The question is not, therefore, whether or not we should use systems analysis or systems concepts in geography, but rather one of examining how we can use such concepts and such modes of analysis to our maximum advantage (Harvey, 1969a, p. 479).

Despite the significant role that Marx's philosophy was to play in

Harvey's thought about three years later, Harvey made only three brief comments on Marx in *Explanation in Geography*. He noted Popper's view, in *The Poverty of Historicism* (1957), that Marx's approach was 'a mode of explanation frequently regarded as deterministic and historicist in the extreme' (Harvey, 1969a, p. 425). Harvey did not dispute this characterisation of Marx's analysis. Earlier, he had referred to the 'mechanistic positivism' of Marx, implying that such a view had application only in very limited circumstances (Harvey, 1969a, p. 47). The only other mention of Marx was a passing reference to his 'dialectical method in history' (Harvey, 1969a, p. 415).

*Explanation in Geography: Methodology and Philosophy.* In turning to Harvey's view of the relationship between methodology and philosophy in *Explanation in Geography*, a distinction is met which Harvey saw to be both 'fundamental' and 'vital' to the book (Harvey, 1969a, pp. 3, 6). He proposed that there were two aspects to geography. There were the 'goals or substantive objects' of study, that is, 'what' was to be studied, or the 'domain' of objects and events to be studied. And, secondly, there was the 'method' of study, that is, 'how' phenomena were to be studied, or the operations of description and explanation (Harvey, 1969a, p. 3). Harvey went on to note that the only grounds on which objectives may 'ultimately' be disputed were grounds of belief. The objectives a geographer chose to pursue were dependent upon that geographer's individual values. These beliefs and values formed the geographer's own 'individual view of life and living' and their manifestation in geography was 'the philosophy of geography'. Thus, as Harvey pointed out, 'defining an objective amounts to assuming a certain philosophical position with respect to geography itself' (Harvey, 1969a, pp. 4-5).

Philosophies of geography have a subjective foundation, but 'all analysis is barren unless there is some objective'. The only objective-less knowledge, suggested Harvey, is the empty analytic understanding provided by mathematical systems and logically constructed calculi (Harvey, 1969a, p. 5). But Harvey did not see it as his purpose to discuss philosophies of geography. He returned to his consideration of the methodology of geography.

> It is ... one of the major tasks of methodological analysis to show how and under what circumstances a particular mode of analysis is appropriate, to specify the assumptions necessary for the employment of a particular technique, and to demonstrate the form of

analysis which must be followed if the analysis itself is to be rigorous and logically sound. All this can be accomplished independently of philosophy (Harvey, 1969a, p. 8).

An explanation may be disputed on logical grounds, although 'some issues regarding explanation cannot be resolved independently of philosophical beliefs', for instance verification and confirmation (Harvey, 1969a, p. 6). Logicians and philosophers, 'particularly the logical positivists and philosophers of science', had examined the logic of explanation and geographers must take account of their criteria for sound explanation. It was the task of the methodologist of geography to consider these criteria in relation to their application to geographical study. The philosopher of geography was concerned with subjective, speculative issues, the methodologist of geography with the objective logic of explanation (Harvey, 1969a, p. 6).

There was a one-way relationship between philosophical and methodological positions, believed Harvey. Philosophers tended to adopt a mode of analysis consistent with their beliefs. This was because such beliefs usually entailed the adoption of a particular method, for example, believing in free will, a philosopher would eschew deterministic models and languages. But, Harvey continued,

the adoption of a methodological position does not entail the adoption of a corresponding philosophical position ... In this respect methodological and philosophical positions are very different from each other ... We cannot argue, therefore, from a methodological position in support of a philosophical position. In the other direction, the relationship is rather closer (Harvey, 1969a, p. 7).

From this point of view, the use of a probabilistic model could not be taken to imply that the researcher had adopted a position of philosophical indeterminacy.

In the concluding chapter of *Explanation in Geography*, Harvey turned his attention to the nature of geography. His aim was to discuss 'how we can use methodology to promote a more concise philosophy of geography' (Harvey, 1969a, p. 481). Noting that formalised theories needed to be given a text to relate them to a particular domain of events, and that un-formalised theories were nevertheless speculations about such a domain, Harvey pointed out that

methodological considerations lead me to conclude ... that it is easiest to identify such a domain (or set of domains) when we possess a well-articulated and well-validated theory . . . The nature of a discipline can be discerned through an examination of the theories which it develops (Harvey, 1969a, p. 483).

For Harvey, the point of view of geography was embodied in geographical theory and geography's subject-matter was to be identified through the text of such theory.

The closest Harvey came to suggesting a definition of geography was in his conclusion that general theory in geography 'will explore the links between indigenous theories of spatial form and derivative theories of temporal process' (Harvey, 1969a, p. 129). He later called this a 'basic tenet of geographical thought' but admitted that 'I find it difficult to state with any certainty the domain of geography' (Harvey, 1969a, p. 483). Harvey did 'speculate' that the 'typical resolution level' at which the geographer studied was neither at the level of patterns of crystals in a snowflake nor stars in the universe but rather at a 'regional' level, although, he believed, it was difficult to pin this down with any precision (Harvey, 1969a, p. 484). Harvey was to conclude that

the closer definition of the domain of geography must wait upon the statement of well-articulated and well-validated geographic theory . . . Without theory we cannot hope for controlled, consistent and rational explanation of events. Without theory we can scarcely claim to know our own identity. It seems to me, therefore, that theory construction on a broad and imaginative scale must be our first priority in the coming decade . . . Perhaps the slogan we should pin up upon our study walls for the 1970s ought to read:
"By our theories you shall know us" (Harvey, 1969a, p. 486).

To depart briefly from the *explication du texte*, it should be noted that there is an ambiguity in Harvey's views in *Explanation in Geography* concerning the relationship between philosophy and methodology. He maintained that the adoption of a methodological position did not entail the adoption of a corresponding philosophical position. However, he also asserted that the philosophy of the scientific method was implicit in quantification. 'If I did not adjust my philosophy, the process of quantification would simply lead me into a cul-de-sac' (Harvey, 1969a, p. vi). Furthermore, Harvey acknowledged his dependence upon the analytic and logical positivist philosophers for the view of the scientific method which he put forward (Harvey, 1969a, pp. 6, 62).

Harvey went on to assert that one ought not to argue from a methodological position in support of a philosophical one (Harvey, 1969a, p. 7). But in his conclusion, as outlined above, he discussed how methodology could be used to promote a more concise philosophy of geography (Harvey, 1969a, pp. 481, 486). Harvey was unable to consistently maintain his initial separation of subjectively-based philosophy from objectively-based methodology. This seriously undermined his argument that the scientific method was an objective standard which should be met by all 'scientific' geographers, irrespective of their philosophy.

One might also ask whether the distinction (and subsequent separation) between philosophy and methodology is a philosophical or methodological one. If it is a philosophical distinction (say, made by analytic philosophy), then it loses any objectivity and is, in fact, given Harvey's view of philosophy, subjective and speculative. It can therefore no longer necessarily be seen to apply to every geographer. On the other hand, if the distinction is a methodological one, then it must be able to be verified logically and objectively. But Harvey's argument for the distinction, as presented in Chapter 1 of *Explanation in Geography*, is based upon the form of a certain definition of geography (Harvey, 1969a, p. 3). The definition has two parts to it, one to do with objectives (and therefore philosophical), the other to do with description and explanation (and therefore methodological), hardly a rigorous, logical and objective basis for a distinction so fundamental to the book.

In *Explanation in Geography*, Harvey drew upon mid-twentieth century philosophy of science to provide a general framework for unified scientific research in geography. This framework was based upon Hempel's deductive-nomological model of scientific explanation and the hypothetico-deductive systematisation of scientific theory. Harvey was able to place the so-called 'quantitative revolution' into perspective, demonstrating the roles of quantification, theory construction, model-building and the development of laws within an overall framework. He captured the spirit of geographical thought at the time and guided it in a certain direction by providing the philosophical and methodological basis for a model-based paradigm, suggesting a strategy, using *a priori* models, for producing scientific laws and theories in geography.

## Behavioural Geography

Harvey's paper, 'Conceptual and Measurement Problems in the Cognitive-behavioral Approach to Location Theory' (1969b), was originally to have been presented at a symposium on behavioural problems in geography at the annual meeting of the Association of American Geographers in

Washington, DC, in August 1968. It was not available at that time but
was published in a collection of the symposium papers in 1969 (Cox
and Golledge, 1969a, pp. iii, 1). In his paper, Harvey was concerned
with the development of a complete decision-making theory which
could be incorporated into location theory. This was a formidable task,
he admitted, but he believed that there were three complementary strategies which could be followed to progress towards such a theory. First,
classical normative location theory could be extended; secondly, a stochastic location theory, that is, a theory based on probability analyses,
could be developed; and, thirdly, a cognitive-behavioural approach could
be constructed (Harvey, 1969b, pp. 37-8). The cognitive-behavioural
approach dealt with 'a "residual" domain of events' (Harvey, 1969b,
p. 40) not able to be analysed by the other approaches. This domain
included, for example, policy problems in which criteria of economic
efficiency were either difficult to define or irrelevant, and the effect
of a small group of individuals upon political and economic decisions.
Thus 'the domain of a cognitive-behavioral location theory will . . .
range from cross-cultural variation in value judgements and perceptions
through individual choice behavior to group decision making processes'
(Harvey, 1969b, p. 40).

Harvey went on to survey the 'conceptual and measurement apparatus' available for the development of a cognitive-behavioural location
theory. He took as 'axiomatic' that 'we can hope to handle the complex
world of behavior only by formulating firm concepts with generally
agreed meanings' (Harvey, 1969b, p. 44). He examined 'economic rationality' and 'satisficing behavior' as examples of 'theoretical concepts'. He
noted that 'economic rationality' had a firm syntactical definition and,
as a primitive term in economic location theory, had been used to generate derived terms such as marginal behaviour and profit maximisation.
'Satisficing behavior', on the other hand, was theoretically ambiguous
and had several connotations, including 'a form of optimizing behavior
in which the criteria are non-economic' and a form of behaviour where
'the decision maker does not seek *any* optimal solution' (Harvey, 1969b,
pp. 44-5). As far as giving 'economic rationality' and 'satisficing behavior' empirical definitions and then measuring them were concerned,
Harvey concluded that the former term had been developed much more
'fruitfully' than the latter (Harvey, 1969b, pp. 45-7).

Noting that some researchers had distinguished between the 'real
world' and the 'perceived world' of the decision-maker, Harvey went
on to examine the concepts and measurement procedures available for
the study of perception. Among others, he considered the potential use

of 'semiotics' that is, the theory of signs. He concluded that, generally speaking, a firmer theoretical framework was needed to facilitate analysis and empirical work in perception studies (Harvey, 1969b, pp. 49-62).

Near the end of the paper, Harvey noted that

> To demand exactness, precision and rigor in the use of terms in the initial stages of investigation can only result in 'the premature closure of our ideas' and hence have a pernicious effect upon the direction of research (Kaplan, 1964, pp. 62-71). Yet we cannot afford to take this as a charter for interminable vagueness in our concepts. Indeed the degree to which we succeed in reducing this vagueness is a measure of our progress (Harvey, 1969b, p. 62).

Geographers needed to gain command over the literature of behavioural science in general or find behavioural scientists interested in geographical problems if a cognitive-behavioural location theory was to be formulated. Harvey concluded that, despite the foundational work by, for example, Isard and Dacey (1962), 'I would doubt if anything very satisfactory will emerge in the way of general theory until the year 2000 A.D. or so' (Harvey, 1969b, p. 64).

Harvey's first published book review was of Pred's (1967) *Behavior and Location*, Part 1, in *Geographical Review* in April, 1969. Pred had 'set out to demonstrate the inadequacies of normative economics', to erect in its place, in geography, a 'behavioral matrix', replacing economic man with behavioural man (Harvey, 1969c, pp. 312-13). Harvey believed that 'it is not a good book . . . It is a golden rule of science that you do not throw away one theory, however inadequate it may be, until you have a better one to replace it' (Harvey, 1969c, p. 313). He also thought that Pred was mistaken in excluding all work on optimisation in regional and location analysis from the geographer's field of interest. Furthermore, Pred's 'behavioral matrix' was based on concepts 'so vaguely defined, so ambiguous, and so completely non-operational' that the idea was 'doomed from the start' (Harvey, 1969c, p. 314). But geographers should not abandon the search for a firm behavioural foundation for their analyses. Harvey believed that Pred had tackled an 'admittedly difficult task' but that the book needed 'much greater rigor and clarity' than that which Pred had achieved (Harvey, 1969c, p. 314).

Both the paper and book review that were published in 1969 reflect Harvey's concern for precision, rigour and clarity in the terms used in behavioural analysis in geography. In the light of the hypothetico-deductive structure of theory that he had examined in *Explanation in Geography*, Harvey was continuing his critique of the then current geographical research even though he realised that any complete or rigorous theory in geography was certain to lie in the distant future.


## From Traditional to Logical Empiricist Geography

The development of Harvey's geographical writings between 1961 and 1969 both reflected and contributed to the growing importance of the spatial science approach in Anglo-American geography during the period. Harvey referred to his time at Cambridge as an 'indoctrination' into 'traditional geography', that is, into the geography associated with Hartshorne's views (Harvey, 1969a, pp. vi, 70-1). He spent the next nine years developing an alternative to the traditional geography of his day by constructing a framework within which he interpreted the role of theories, laws and models in scientific geographical explanations. *Explanation in Geography* represents the consolidation and integration of Harvey's contribution to the development of this alternative approach.

The philosophical viewpoint apparent in Harvey's writing since at least 1966 was a combination of what might be called logical empiricism and instrumentalism, terms which Harvey himself did not use during this period. Harvey relied upon the views of Carnap, Hempel, Nagel and Braithwaite who had developed the deductive-nomological model of scientific explanation and the hypothetico-deductive structure of scientific theory. Brown has noted that these four philosophers were central to the logical empiricist movement in the philosophy of science (Brown, 1977, pp. 23, 25, 38, 63). Logical empiricism was a development and modification of the original logical positivist programme (Joergensen, 1951, p. 40; Brown, 1977, p. 23). The central difficulty for logical positivism as a philosophy of science was that scientific laws formulated as universal propositions could not be conclusively verified by any finite set of observation statements (Brown, 1977, p. 23). Most of the logical positivists gave up such a strict definition of scientific laws and replaced it with the requirement that

a meaningful proposition must be testable by reference to observation and experiment. The result of these tests need not be conclusive, but they must provide the sole ground for determining the truth or falsity of scientific propositions ... Rudolph Carnap's *Testability and Meaning* [Carnap, 1936, 1937] can reasonably be viewed as the founding document of logical empiricism (Brown, 1977, p. 23).

The 'logical' aspect of logical empiricism refers to its emphasis upon the logical structure of explanation and theory, that is, the modelling of what is 'scientific' upon the logical structure of Euclid's geometry and the supposed logic of explanation in physics. The 'empiricist' aspect refers to the view that all human experience is built up from unordered sense perceptions, that is, bits of 'information' received by the 'physical' senses, and that scientific laws are to be tested against such sense perceptions.

Hempel's model of explanation was 'the classic statement' of the deductive pattern of scientific explanation generally accepted by logical empiricists (Brown, 1977, p. 51; Benton, 1977, pp. 47-61) while the hypothetico-deductive account of scientific theories has been described as 'widely accepted' among logical empiricists (Benton, 1977, pp. 64-7). However, Suppe, in his extensive exposition of the logical empiricist view of theories, which he called the 'Received View', noted that by the late 1960s, the logical empiricists' views had been widely acknowledged by philosophers of science to be inadequate (Suppe, 1974, p. 4). Alternative approaches had been advocated by such philosophers of science as Toulmin (1953), Kuhn (1962) and Feyerabend (1963), who emphasised the role of historical and social factors in scientific understanding, although no one new approach had been generally accepted. This means that Harvey was presenting a view of science that had dominated the philosophy of science for more than forty years but which was then coming under increasing criticism.

In his publications, Harvey stated that the view of science he was presenting was that of 'philosophers and logicians' (Harvey, 1967b, p. 4, 1967c, pp. 211, 213), of 'modern analytic philosophy' (Harvey, 1967c, p. 211), of 'the philosophers of science' (Harvey, 1969a, p. 62) and of 'logical positivists' (Harvey, 1969a, p. 6). There is often no distinction made in the philosophy of science between logical positivism and logical empiricism and, as Caponigri has pointed out,

The dominant current of philosophical thought in the English-speaking world today is called by a variety of names: neo-empiricism, logical positivism, logical empiricism, analytic philosophy and linguistic analysis . . . Through the entire range of these movements there runs an element of unity . . . , the view that the main concern of philosophy is the analysis of language (Caponigri, 1971, p. 301).

The work of the logical empiricists upon which Harvey relied clearly illustrates this concern with 'the analysis of language' in the philosophy of science, exemplified in the hypothetico-deductive account of the structure of theory.

The emphasis on rationality, order and form in logical empiricism is reflected in the wood engraving 'Other World' by M.C. Escher reproduced on the cover of *Explanation in Geography*. Here, viewers find themselves in a room looking out on a lunar landscape through three sets of two windows, each set of windows having a Persian man-bird and a lamp in it (Ernst, 1976, pp. 46-7). One set of windows is seen from the side, another from above and the third from below. There is a fascinating orderedness and symmetry about the scene which mirrors Harvey's fascination with the logical empiricist scientific method and its 'fantastic power' (Harvey, 1969a, p. vi). But as it becomes apparent that the world which Escher has constructed cannot be real, so the question may be posed whether the world reconstructed via this scientific method is also real.

Harvey's rejection of 'logical positivists of the extreme variety, who have held that all knowledge and understanding can be developed independently of philosophical presuppositions' (Harvey, 1969a, p. 8), was a rejection of that trend in logical positivism which regarded metaphysical statements as meaningless (Joergensen, 1951, pp. 4-5). Harvey viewed metaphysical speculations about the nature of geography as indispensable in providing direction to theory construction (Harvey, 1969a, p. 114).

In surveying the philosophy of science literature, Harvey noted some of the views of Popper and Kuhn. Popper was closely associated with logical empiricism but called himself a 'critical rationalist' (Popper, 1976, p. 116). Harvey reviewed Popper's ideas on falsification in *Explanation in Geography* (Harvey, 1969a, p. 39). Kuhn's (1962) account of scientific revolutions was one of the works which prompted Harvey to briefly consider science as an activity in *Explanation in Geography* and to note that the 'analytic' approach regarded experience as but a set of so-called 'factual' statements about reality (Harvey, 1969a, pp. 9-10). But Harvey's reservations about logical empiricist philosophy of science did not effectively go beyond these remarks. His strategy for the construction of scientific theory assumed the validity of both the deductive-nomological model of explanation and the hypothetico-deductive structure of scientific theory (Harvey, 1969a). His assessment of the concepts of the economic and behavioural approaches to location theory was based upon the criteria set by the logical empiricist view of science (Harvey, 1969b). Thus, Harvey aligned himself with the essential aspects of logical empiricist philosophy of science.

Central to Harvey's strategy for the development of laws and the construction of theories in geography was the 'willingness to regard events *as if* they are subject to explanation by laws' (Harvey, 1969a, p. 174). Furthermore, using *a priori* models to help develop theory (Harvey, 1969a, p. 175) was effectively to view such models *as if* they were theories. Almost ten years later, Gregory was to point out that injunctions using 'as if' in this way 'flow from an instrumentalist conception of science' (Gregory, 1978, p. 40). Keat and Urry have noted that instrumentalism in this sense denoted a view about the logical status of scientific theories, namely,

> that they are computational devices which generate testable predictions. Theories are instruments and, as such, only their utility can be assessed, and not their truth or falsity. They do not provide any knowledge of the physical world over and above the predictions that can be derived from them (Keat and Urry, 1975, p. 63).

For Harvey, theories do provide true explanations but geographers' 'lack of understanding' about theory construction and the 'extreme complexity' of geography's subject-matter (Harvey, 1969a, p. 173) meant that no complete geographical theories existed and that their development would be extremely difficult and would take some time. Harvey was therefore forced to enunciate a strategy that in effect involved the replacement of theories with models that could generate testable hypotheses and predictions. In doing so, this strategy resorted to an instrumentalist conception of science although it was based upon the logical empiricist philosophy of science.

Harvey rejected Hartshorne's idiographic method, advocating instead a nomothetic approach central to which was the development and use of generalisations and laws. He also rejected Hartshorne's definition of geography. Even in his doctoral thesis, Harvey noted that the consideration of economic processes took his work out of the sphere of geography 'from Hartshorne's standpoint' (Harvey, 1961, p. ii). Later, in *Explanation in Geography*, Harvey looked to the development of scientific theory in geography to more closely delimit the domain of the discipline, although he was of the view that geography focused on the spatial form of phenomena at the 'regional level' (Harvey, 1969a, p. 484).

But it was necessary for geography to deal with process as well as form. Harvey's concern with the relationship between form and process in geography is apparent throughout the 1960s. Without understanding processes, geographers could not hope to explain the dynamics of locational change. He concluded that geographers were concerned with 'spatial processes', that is, 'spatial manifestations of temporal phenomena' (Harvey, 1967b, p. 7, 1967c, p. 213). Geography depended upon two sets of theoretical postulates, indigenous postulates relating to spatial form and derivative postulates relating to temporal processes (Harvey, 1967b, pp. 7-9, 1967c, p. 213, 1969a, pp. 117-27). Geographers therefore needed to become more familiar with other disciplines in order to gain an understanding of the processes they investigated. Harvey thus entered the beginning of a new decade, the 1970s, with an interdisciplinary attitude towards scholarship, seemingly poised to continue his contribution towards the development of rigorous, scientific theory in geography.

## Notes

1. The following account of the development of spatial science geography is drawn largely from James (1972) and Johnston (1979).

2. This is not to deny the existence of significant alternative approaches, for example, that of Sauer, who emphasised the cultural landscape and its development (Sauer, 1925; James, 1972, pp. 399-404; Leighly, 1979, pp. 4-9).

3. Harvey's doctoral supervisor was C.T. Smith, an historical geographer (Smith, 1965, 1967).

4. These remarks (Harvey, 1970c) are addressed to Harvey's (1963) first published article which included a number of the correlation and regression analyses initially contained in his doctoral thesis. Extending the remarks to the thesis is thus not considered to be misleading.

5. The definitions of 'normative', 'analytic' and 'descriptive' models used here were not included in Harvey's (1967b) seminar paper but were inserted in the 1970 version, which was published in *Geographia Polonica*.

6. Harvey gives as an example of an axiomatic system in geography that of Garrison and Marble (1957), which dealt with the spatial structure of agricultural activity (Harvey, 1969a, pp. 132-4).

7. This means that models may be used at every step from 'theory' to 'verification' as set out in Figure 2.4.

# 3 EMBRACING MARXIST METHOD: FROM SOCIAL JUSTICE TO OPERATIONAL STRUCTURALISM

In 1973, *Social Justice and the City*, Harvey's second book, was published. It contained six essays which had been written over the previous three to four years during which time Harvey had embraced a Marxist approach to geography. In the first three essays, he explored the possibility of incorporating social, political and ethical objectives such as social justice in theories used in urban and territorial planning. In the process of this exploration, Harvey became dissatisfied with both the particular theoretical approach of logical empiricism and with the unjust character of capitalist society. Subsequently, in the last three essays in *Social Justice and the City*, Harvey adopted Marx's conceptual and methodological approach. In the seventh, and concluding, chapter of the book, Harvey characterised Marx's method as based upon an operational structuralist view of society. Thus the essays making up *Social Justice and the City*, the first four of which had previously been published elsewhere, represent the evolution of Harvey's work from that utilising a liberal approach concerned with social justice to that which embodied a socialist perspective based upon Marx's operational structuralist method. In further publications in 1973 and 1974, Harvey attacked the supposed neutrality of the scientific method of logical empiricism, contrasting it with the committed and more critical method of Marx.

## Relevance, Radicalism and the Geography of the Early 1970s

In 1971, Chisholm sparked off a debate on relevance in geography in *Area*, a journal of the Institute of British Geographers.[1] He identified as an 'emerging weakness' of geography its 'lack of a corpus of normative statements and ideas'. Without this, geography could not offer its students, nor the governments of the world, any suggestions for improving the world (Chisholm, 1971). Later in the same year, Prince and Smith reported upon the 1971 annual meeting of the Association of American Geographers. Smith noted that 'a new wind of change is beginning to blow, in the form of ... a movement ... calling for a greater professional involvement with matters of contemporary social concern' (Smith, 1971, p. 153). Berry discerned two main groups in this movement. One consisted of 'white liberals, . . . bleeding hearts' who were quick to avoid hard work. The other group was made up of 'hard-line Marxists'. Neither was committed 'to producing constructive change by democratic means' (Berry, 1972a, p. 77). Berry suggested that geographers become part of society's decision-making apparatus in order to effect change. Blowers (1972, p. 291) saw Berry's suggestion as threatening the political independence of geographers' contributions. As the debate progressed, positivistic geography was criticised for its 'unattainable optima' (Smith, 1973b, p. 3), its 'straitjacket of neo-classical economic assumptions' (Eyles, 1973, p. 159; Gray, 1975) and its 'feeble' impact upon the planning process (Blowers, 1974, p. 32).

The issue of relevance in geography was carried forward by the McGraw-Hill Problems Series in Geography, a set of seven books edited by E.J. Taaffe. They were introduced on their covers as 'provocative examples of the application of geographic research to selected contemporary problems involving the city, environmental quality, and regional development' (Rose, 1971; Morrill and Wohlenberg, 1971; Bach, 1972; Smith, 1973a; Cox, 1973; Harries, 1974; Shannon and Dever, 1974). In 1973, Albaum edited a reader entitled *Geography and Contemporary Issues: Studies of Relevant Problems*, a collection of 38 articles arranged around the themes of poverty, black America, urban housing, environmental issues, population growth and pressure, and conflict and conflict resolution. Harvey's article, 'Social Justice and Spatial Systems', was reprinted in the last section of the book.

Also during this period, Olsson (1970) and Gale (1972b) had been advocating what Smith has called a 'logico-linguistic' approach (Smith, 1979, p. 361). Retaining the emphasis on the importance of theoretical languages which had characterised spatial science geography, they explored the use of 'many-valued logics' and 'fuzzy sets' and their potential to provide 'more realistic accounts of the ambiguities inherent in human values, decision making, and action'. Their approach was directly geared towards the 'manipulation of society and its various components' in the human interest (Olsson, 1970, p. 370).

The radicalism that Smith (1971) had noted as being evident at the annual meeting of the Association of American Geographers in 1971 had its origins at the end of the 1960s. In 1968, for instance, Bunge founded the Society for Human Exploration, calling for geographers to 'explore' the poorest and most blighted areas of North America, planning with people, not for them, in an attempt to overcome the elitist professionalism of academic geography (Bunge, 1969, 1971). In 1969,

*Antipode*, 'A Journal of Radical Geography', was founded at Clark University, Massachusetts. Its aim, as stated in the first editorial, was 'to ask value questions within geography, question existing institutions concerning their rates and qualities of change, and question the individual concerning his own commitments' (Wisner, 1969, p. iii). The radical geography of *Antipode*'s early years took a 'liberal' form, according to Peet (1977b, p. 249), with interest focused upon the relevance issue. The 'breakthrough to Marxism' was not achieved until the publication of Harvey's 'Revolutionary and Counter-revolutionary Theory in Geography and the Problem of Ghetto Formation' in 1972 (Peet, 1977b, p. 249). From that time onwards, many radical geographers began to use Marxist theory in attempting to construct 'a radical philosophical and theoretical base for a socially and politically engaged discipline' (Peet, 1977b, p. 250).

Two further challenges to spatial science geography in the early 1970s also drew upon previously well-established philosophical systems, namely, phenomenology and idealism. The first article on phenomenology appeared in the *Canadian Geographer*, attacking the philosophical and methodological dominance of positivist approaches in geography (Relph, 1970). Relph pointed out 'the dictatorship and absolutism of scientific thought over other forms of thinking' that was characteristic of positivism and noted that phenomenology attempted to formulate a method of investigation alternative to that of 'hypothesis testing and the development of theory'. Phenomenology's method, the outline of which Relph took from Spiegelberg (1960), centred upon careful description and examination of 'the entire structure of the phenomena being studied in all its possible meanings' (Relph, 1970, pp. 193-4). Two other articles on phenomenology appeared in the *Canadian Geographer* in the early 1970s (Tuan, 1971; Walmsley, 1974).

In 1971, also in the *Canadian Geographer*, Guelke attacked as 'narrow' and 'one-sided' both Hempel's deductive-nomological model of scientific explanation and the view of the role of laws, theories and models in geography that Harvey (1969a) and Chorley and Haggett (1967), among others, had been advocating. Guelke believed that this type of geography had not yet produced any scientific laws and that the theories and models that had been produced were not amenable to empirical testing (Guelke, 1971, pp. 50-1). Guelke later went on to advocate an idealist approach in human geography which explained rational human action by reconstructing the thought behind it rather than by reference to laws or theories (Guelke, 1974).

Harvey's publications over this period had initially much in common with those struggling with the issue of relevance in geography. But Harvey quickly took up a radical stance in preference to logico-linguistic, phenomenological and idealist options.

## David Harvey's Geography, 1970-1974

In 1970, Harvey had moved to Johns Hopkins University in Baltimore and had begun research into some philosophical aspects of geographical theory. Three publications resulted, exploring the incorporation of political and ethical objectives in urban theory (Harvey, 1970a, 1971, 1972b). A fourth article heralded Harvey's adoption of a Marxist perspective (Harvey, 1972c). The transition to Marxism is also reflected in the views on the nature of theory put forward by Harvey during this period (Harvey, 1970c, 1972a, 1972f). Harvey took part in research on the Baltimore housing market (Harvey, Chatterjee, Wolman, Klugman and Newman, 1972) which he incorporated in a paper on urbanism written for the Association of American Geographers' Commission on College Geography (Harvey, 1972g). The tentative Marxist formulations of this paper gave way to a more thoroughly Marxist analysis in the essays which were published as Chapters 5 and 6 of *Social Justice and the City*. In Chapter 7 of that book, and in four further publications, Harvey explored the nature of Marx's method and compared it with the methods of spatial science geography (Harvey, 1973b, 1974a, 1974c and 1974d).

### Spatial Processes and Social Justice

At a meeting in November 1969, of the Regional Science Association at Santa Monica, California, Harvey presented a paper entitled 'Social Processes and Spatial Form: An Analysis of the Conceptual Problems of Urban Planning' (Harvey, 1970a). A second paper, 'Social Processes, Spatial Form and the Redistribution of Real Income in an Urban System' was initially included in the first but 'cut out because of its length' (Harvey, 1973a, p. 18). It was presented at the Twenty-second Symposium of the Colston Research Society, held at the University of Bristol in April 1970, and was published in 1971 (Harvey, 1971). In both of these papers, Harvey continued to deal with problems associated with the relationship between social processes and spatial form but now within the context of the city and urban planning. The specific issues he examined were those of the relationship between sociology and geography (Harvey, 1970a) and planning for the redistribution of income through the provision of public goods (Harvey, 1971).

In the first paper, Harvey, relying on Mills (1959), characterised the 'sociological imagination' as dealing with 'history and biography and the relations between the two in society' (Mills, 1959, p. 5) whereas he identified the 'geographical imagination' with a 'spatial consciousness' (Harvey, 1970a, p. 48). Noting that an adequate conceptual framework for understanding the city must encompass both disciplines but that the two rarely interacted, Harvey explored 'the philosophy of social space' as an example of where interaction was important. He concluded that 'social space is complex, nonhomogeneous, perhaps discontinuous, and almost certainly different from the physical space in which the engineer and the planner typically work' (Harvey, 1970a, p. 58). Understanding social space thus depended upon understanding social processes and was therefore a step towards the integration of the sociological and geographical imaginations. But the gap between the two imaginations could be bridged only if adequate tools were available.

> These tools amount to a set of concepts and techniques which can be used to weld the two sides together. If the resultant construct is to be capable of analytic elaboration and susceptible to empirical testing, then mathematical and statistical methods will be needed . . . It seems likely that these methods will not be identifiable except in a given context . . . I cannot, therefore, set up any general methodological framework for working at the interface (Harvey, 1970a, pp. 59-60).

Harvey went on to examine three problems associated with inference and predictive control when analysing the gross spatial form of a city and aggregate overt behaviour patterns within it. The first problem was that geography and sociology used two different types of language to define a set of individuals. In order not to confuse these languages, a 'metalanguage', embracing both, could be used, although Harvey was quick to note that its development would not be easy (Harvey, 1970a, p. 61). The second problem was that of 'confounding' where, for instance, researchers would confound causes with effects and functional with causal relationships (Harvey, 1970a, pp. 62-3). The third problem was associated with the use of statistical inference. Harvey pointed out that there was no accepted definition of statistical significance in spatial inference (Harvey, 1970a, p. 63). He felt forced to conclude that

> we do not possess very sharp tools for dissecting the problems which

arise when we seek to blend together sociological and geographical techniques to deal with urban problems. Consequently, we must anticipate difficulty in making conditional predictions and in validating theory (Harvey, 1970a, p. 65).

Harvey decided that 'temporary frameworks with which to construct a theory of the city' would have to be devised (Harvey, 1970a, p. 65). His 'strategy' was based upon regarding the city as a 'complex dynamic system in which spatial form and social processes are in continuous interaction with each other', avoiding deterministic views giving priority to either form or process (Harvey, 1970a, pp. 65-7). Harvey's strategy was similar to that developed in an earlier article (Harvey, 1968b), that is, it was an 'iterative' one,

> in which we move from spatial form manipulation (with social process held constant) to the social process implications (with the new spatial form held constant). We can move in either direction . . . This seems to be the style that is developing in urban modelling (Harvey, 1970a, p. 68).

In the second of these two papers, Harvey reflected upon the failure to develop an overall strategy, for dealing with urban systems, which reconciled policies designed to change the spatial form of a city with policies concerned to affect urban social processes in order to achieve some 'coherent social objective'. The complexity of the city, the myopia of a disciplinary approach and the difficulty of defining a coherent social objective all contributed to this failure. As far as a coherent social objective for social planning and forecasting was concerned,

> it involves a set of social, political and ethical judgments upon which it will be very difficult to obtain general agreement. The trouble with merely dodging the issue is that judgments are inevitably *implied* by a decision, whether we like it or not. If, for example, we predict as well as we can on the basis of current knowledge and trends the future population distribution, consumption patterns, travel demands (by mode) and so on, and allocate current investment accordingly, we thereby imply that these future conditions are acceptable to us (Harvey, 1971, p. 267).

Harvey suggested that the ethical judgement central to any social policy formulation with respect to a city system was that of how much income redistribution should take place. Stating that 'I am generally in favour of a far more egalitarian social structure than currently exists in either American or British urban systems', Harvey went on to demonstrate how income inequalities were usually increased by the 'hidden mechanisms' of income redistribution. He defined income as 'an individual's command over society's scarce resources' (Harvey, 1971, p. 269) although it was later noted that cultural, religious, ethnic, class and other factors meant that 'resources mean different things to different people' (Harvey, 1971, p. 291).

Harvey showed how the 'price of accessibility' to employment, resources, welfare services and so on, and the 'cost of proximity' to sources of pollution and noise, for instance, would be affected by any change in the spatial form of the city. The change, in turn, would bring about redistributions in real income (Harvey, 1971, p. 272). Harvey believed that the market mechanism was unable to allocate resources efficiently when externalities, 'unpriced and nonmonetary effects', were present (Harvey, 1971, pp. 272-4). Groups with greater resources, for example money, influence and information, would dominate political locational decisions at the expense of the usually larger poorer and less-educated groups. Quoting Sherrard (1968, p. 10), Harvey observed that

> The slum is the catch-all for the losers, and in the competitive struggle for the cities' goods the slum areas are also the losers in terms of schools, jobs, garbage collection, street lighting, libraries, social services, and whatever else is communally available but always in short supply. The slum, then, is an area where the population lacks resources to compete successfully and where collectively it lacks control over the channels through which such resources are distributed or maintained (Harvey, 1971, p. 287).

In the concluding section of the paper, Harvey noted that mechanisms governing the redistribution of income seemed to be moving the cities of Britain and the United States towards a state of greater inequality, greater injustice and potential violent conflict.

> I therefore conclude that it will be disastrous for the future of the social system to plan ahead to facilitate existing trends — this has been the crucial planning mistake of the 1960's ... An enormous task confronts us. We really do not have the kind of understanding of the total city system to be able to make wise policy decisions,

even when motivated by the highest social objectives. It seems, therefore, that the formation of adequate policies and the forecasting of
their implications is going to depend for their success upon some
broad interdisciplinary attack upon the social process and spatial
form aspects of the city system (Harvey, 1971, p. 298).

In his first paper, Harvey emphasised the role of language and of
mathematical and statistical methods, that is, what might be called the
'form' of theoretical investigation. In his second paper, he showed far
more concern for the issue of income redistribution as such, that is,
with what might be called the 'content' of theoretical investigation.
Harvey had turned his attention to the plight of those disadvantaged
by the outworkings of what he called the 'market mechanism'. He expressed a dissatisfaction with the *status quo*. At this stage, his solution
to the problems of the city in Britain and the United States appeared
to be one of an interdisciplinary approach to urban forecasting and
planning.

At the 1971 annual meeting of the Association of American Geographers, there was held a special session entitled 'Geographical Perspectives on American Poverty and Social Well-being', reflecting the
growing concern for issues of contemporary relevance. At this session
Harvey presented a paper entitled 'Social Justice and Spatial Systems'
in which he extended his previous analyses by exploring 'the possibility
of constructing a normative theory of spatial or territorial allocation
based on principles of social justice' (Harvey, 1972b, p. 88). In examining how a 'just distribution' may be 'justly arrived at', Harvey stated
that he assumed that it was possible to devise a socially just definition
of income and that 'justice achieved at a territorial level of analysis
implies individual justice even though I am all too aware that this is not
necessarily the case' (Harvey, 1972b, pp. 89-90).

Harvey considered the problem of a central authority seeking to allocate scarce resources over a set of territories in such a way that social
justice was maximised. He believed that if 'need', 'contribution to
common good' and 'merit' could each be evaluated and measured for
the territories, arriving at a hypothetical figure for the allocation of
resources among them, then it would be possible to evaluate existing
distributions and devise policies to improve them (Harvey, 1972b,
p. 91). Harvey outlined the meaning of need, contribution to common
good and merit in a territorial context, noting problems of definition
and measurement (Harvey, 1972b, pp. 92-6). He concluded that the
concrete application of such criteria would be very difficult but that

they provided the beginnings of 'a normative theory of spatial organization based on territorial distributive justice' (Harvey, 1972b, p. 96).

When dealing with 'justly arriving at' an income distribution, Harvey
noted that both Marx and the constitutional democrats assumed that
'if socially just mechanisms can be devised then questions of achieving
social justice in distribution will look after themselves' (Harvey, 1972b,
p. 97). The failure of the socialist programmes of post-war Britain and
the liberal anti-poverty programmes in the United States was due to the
fact that they sought 'to alter distribution without altering the capital-
istic market structure within which income and wealth are generated
and distributed'. Under an 'individualistic capitalist system', it was
accepted as 'rational and good' for capital to flow to wherever the rate
of return was highest, creating localised pockets of 'high unfulfilled
need' (Harvey, 1972b, pp. 98-9). Harvey outlined the example of the
private inner-city rental housing market which had collapsed as capital
had been attracted to the suburbs.

> If we bribe financial institutions to get back into the market, one
> effect will be to create a greater relative scarcity of capital funds for
> (say) suburban development. The more advantaged suburbs will
> adjust the rate of return they offer upward to bring back the capital
> flow. The net effect of this process will be a rise in the overall rates
> of return which is obviously to the advantage of financial institu-
> tions . . . What this suggests is that there is a built-in tendency for
> the capitalist market system to counteract any attempt to divert the
> flow of funds away from the most profitable territories . . . [and]
> that "capitalist means invariably serve their own, capitalist, ends"
> [Huberman and Sweezy, 1969] , and . . . these capitalist ends are not
> consistent with the objectives of social justice (Harvey, 1972b,
> p. 100).

The capitalist system depended upon scarcity to define exchange values
and it was the production and maintenance of scarcity which led to
deprivation, appropriation and exploitation (Harvey, 1972b, p. 101).
Harvey admitted that the capitalist market system increased the total
product available to society but that the need to continue scarcity led
to 'socially undesirable' consumption, for example, militarism and con-
spicuous consumption. He pointed to China and Cuba as significant
experiments in promoting 'growth with social justice' (Harvey, 1972b,
p. 102).

Harvey concluded that if territorial social justice was to be achieved, then the institutional, organisational, political and economic mechanisms of society should be such that the prospects of the least advantaged territory were as great as they possibly could be. Furthermore, questions of social justice should not be ignored 'for to do so amounts to one of those strategic non-decisions . . . by which we achieve a tacit endorsement of the *status quo*. Not to decide on these issues is to decide' (Harvey, 1972b, p. 104).

In explicitly asking value questions within the context of location theory, Harvey went beyond the analyses of the first two papers. He formulated the view that, in avoiding ethical questions, a neutral stance did not result. Rather, the *status quo* was upheld. And the *status quo* of contemporary capitalism was, for Harvey, quite undesirable because of its apparent injustices and inequalities.

### Marxist Analysis and Socialist Solutions

Harvey's 'Revolutionary and Counter-revolutionary Theory in Geography and the Problem of Ghetto Formation' was published in *Geography of the Ghetto: Perceptions, Problems and Alternatives*, Volume 2 of the 'Perspectives in Geography' series (Rose, 1972). Harvey began by posing the question 'How and why would we bring about a revolution in geographic thought?' He pointed out that Kuhn's (1962) view of scientific revolutions provided no explanation as to how anomalies arose and generated crises, nor how a new paradigm became accepted. This was because Kuhn's analysis was 'idealist', abstracting scientific knowledge from its 'materialistic base' (Harvey, 1972c, p. 4). Relying on Johnson (1971), Harvey outlined five characteristics that a new theory had to possess in order to be accepted. It had to attack the central proposition of conservative orthodoxy, appear to be new but absorb much of the orthodox theory, be difficult to understand in order to exclude senior academics and challenge younger ones who could nevertheless master it, appear to offer a new methodology and offer an important empirical relationship to measure (Harvey, 1972c, pp. 5-6). 'The history of geographic thought in the last ten years is exactly mirrored in this analysis', Harvey pointed out.

> The quantitative movement can thus be interpreted partly in terms of a challenging new set of ideas to be answered, partly as a rather shabby struggle for power and status within a disciplinary framework, and partly as a response to outside pressures to discover the means for manipulation and control in what may broadly be defined as "the planning field" (Harvey, 1972c, p. 6).

Harvey went on to define a counter-revolutionary theory as one deliberately proposed to deal with a revolutionary theory so that 'the threatened social changes which general acceptance of the revolutionary theory would generate' would be prevented from being realised. The revolutionary co-optation of Marxist theory in Russia after Lenin's death and in Western sociology without conveying the 'essence of Marxist thinking' effectively prevented the 'true flowering' of Marxist thought and, concomitantly, the emergence of the 'humanistic society' which Marx had envisaged. Harvey proposed replacing 'manipulation and control' with 'the realization of human potential' as the basic criterion for paradigm acceptance (Harvey, 1972c, pp. 7-9).

Having explored how revolutions in geography might occur as well as some of their possible implications, Harvey went on to look at why a revolution was needed. He concluded that the 'sophisticated theoretical and methodological framework' of the 'quantifiers' could no longer say anything meaningful about such events as the ecological, urban and international trade problems which were unfolding around them (Harvey, 1972c, p. 10).

In accomplishing the required revolution, Harvey rejected 'abstract idealism' as lacking any real content and while phenomenology viewed people as being in 'constant sensuous interaction with the social and natural realities' surrounding them, it could easily lead into idealism or back into 'naive positivist empiricism'. Harvey identified Marxist thought as 'a socially aware form of materialism' and as 'the most fruitful strategy' in exploring the overlap of positivism, materialism and phenomenology (Harvey, 1972c, pp. 10-11). Positivism sought to understand the world, restricting itself to an analysis of a defective existing reality, whereas Marxism sought to change the world. The dialectical method of Marxism allowed concepts and categories to change with a changing reality. Harvey pointed out that Marx had derived his phenomenological position and dialectical method from Hegel (Harvey, 1972c, p. 11). Harvey thus rejected the logico-linguistic, phenomenological and idealist options which were being offered in contemporary geography. He was directing the concern for relevance towards the development of a radical Marxist perspective.

Turning his attention to the question of understanding and solving the ghetto problem, Harvey believed that 'liberal' solutions sought to cure inequity within an existing set of social mechanisms whereas a 'revolutionary' solution would eliminate those mechanisms. As a simple example, Harvey assumed as valid the positivist theory, developed by Muth (1969) from Alonso (1964), that competitive bidding determined urban land-use and therefore ghetto formation. Harvey suggested that a socially controlled urban land market and socialised control of the housing sector should replace the mechanism of competitive bidding, as had occurred in Cuba, and the problem of ghetto formation should then be solved.

> This argument with respect to the Alonso-Muth residential land-use theory is overly simplistic. Since it is frequently the case that a mechanism which is assumed for the purposes of the theory is not necessarily the same as the real mechanisms which generate results in accordance with the theory, it would be dangerous indeed to point immediately to competitive market processes as being the root cause of ghetto formation (Harvey, 1972c, p. 18).

Harvey concluded that geographers needed to use their 'powers of thought' to formulate concepts and theories that could be applied to bringing about 'a humanizing social change' (Harvey, 1972c, p. 24).

> The emergence of a true revolution in geographic thought is bound to be tempered by commitment to revolutionary practice . . . If conditions are as serious as many of us believe, then we will increasingly come to recognize that nothing much can be lost by that kind of commitment and that almost everything stands to be gained should we make it and succeed (Harvey, 1972c, p. 25).

Owing to its pronounced turn to Marxist analysis, Harvey's 'Revolutionary and Counter-revolutionary Theory in Geography and the Problem of Ghetto Formation' was also published in *Antipode* in 1972 (Harvey, 1972d). There followed six comments on the paper. Folke (1972) called it a 'pioneering article' in that it acknowledged the consequences of capitalism for geography and for the ghetto. He also emphasised the close relationship between revolutionary theory and practice.

> It is crucially important to understand that theory cannot be developed first and then put into practice. The revolutionary process must be a dialectical one between theory and practice. Revolutionary theory without revolutionary practice is not only useless, it is inconceivable (Folke, 1972, p. 17).

Hayford commented that Harvey had 'made the beginnings of a good

critique of the actual situation of geography as an academic discipline
and of geographers as social scientists' but she too asked for clarifica-
tion of the question of the relationship between revolutionary theory
and practice (Hayford, 1972, pp. 20-1). Campbell (1972) viewed Marxist
solutions as unrealistic and advocated a combination of positivist and
existentialist approaches whereas Olsson (1972), noting that he shared
most of Harvey's intentions, disagreed with his prescriptions and his
emphasis upon only one particular approach to scientific problems.
Berry agreed with Harvey's condemnation of 'the academic bleeding-
heart within geography' who was characterised by moral indignation
and trivial research. However, he referred to Harvey as a 'radical liberal'
who 'remains a product of the white European experience; he still has
faith in logical rationalism, and views the central city negatively as a
problem' (Berry, 1972b, p. 33). Berry also accused Harvey of wishful
thinking in believing that the development of Marxist theory would
automatically or 'magically' lead to social change (Berry, 1972b, p. 32).
Finally, Getis (1972) agreed with Harvey that a new paradigm was
needed for geography but he was not satisfied that Marxism was the
only way to achieve this. Harvey's analysis was at an 'institutional scale'
while work could also be carried out on the 'psychobiological level'.

In his 'Commentary on the Comments', Harvey discussed two issues,
namely, the market process and the relationship between social and
disciplinary revolutions (Harvey, 1972e). He accepted the view of Marx
and Engels (1971) that the ruling class produced the ruling ideas and
organised the knowledge of society.

> This means that *in general* all knowledge is suffused with apologetics
> for the *status quo* and counter-revolutionary formulations which
> function to frustrate change . . . The pursuit of knowledge and the
> organization of knowledge is inherently conservative (a position
> which the general spread of scientific method of the logical-empiricist
> sort has done much to reinforce) (Harvey, 1972e, p. 39).

All disciplinary boundaries were counter-revolutionary, according to
Harvey, because they allowed the State to 'divide and rule'. Reality
had to be approached directly and conceptualised in 'non- or meta-
disciplinary terms' (Harvey, 1972e, p. 40). Geographers had to start
where they were and question and reformulate geographical theory.
*Status quo* theory accurately described reality at a particular point in
time and its prescriptive policy solutions perpetuated the *status quo*.
Counter-revolutionary theory generally obscured an understanding of

reality whereas revolutionary theory, recognising that reality is a dialectical process, formulated propositions as only contingently true and identified 'real' future choices in the present social situation. The implementation of revolutionary choices validated the theory and provided grounds for the formulation of a new theory. According to Harvey, any theoretical formulation could become any one of these three types of theory (Harvey, 1972e, p. 41).

In 1973, Morrill also commented upon 'Revolutionary and Counter-revolutionary Theory in Geography and the Problem of Ghetto Formation' (Morrill, 1973). In articles previously published in *Antipode*, he had rejected Marxist solutions to social problems (Morrill, 1969, 1970). In his 1973 paper, he asserted that geometric theories of optimal location were not inherently counter-revolutionary. Harvey, in reply, agreed, although he pointed out that 'location theory . . . has no meaning independently of our concepts of rent, capital and labor' (Harvey, 1973c, p. 87). The source of geography's location theory, namely, neo-classical marginalist economics, was an inadequate conceptual framework because it defined capital independently of its relation to labour (Harvey, 1973c, p. 87).

In 'Social Justice and Spatial Systems' and 'Revolutionary and Counter-revolutionary Theory in Geography and the Problem of Ghetto Formation', Harvey, for the first time in publication, clearly stated his dissatisfaction with the capitalist market economy in its distribution of society's wealth. In turning to Marx and Engels, he discovered a more satisfactory and 'realistic' line of approach than that of orthodox approaches. Marxism was no longer viewed as a 'mechanistic determinism', as in *Explanation in Geography*. It was rather a viewpoint seeking to transform a defective reality by the formulation of revolutionary theory and practice. Harvey's consideration of Marxism led him to replace questions of distribution with questions of production and the creation of surplus value. Furthermore, Harvey's attention had turned from the logical empiricist view of the internal structure of scientific theory to the external role of theory, that is, to the role of theory in relation to society, as revolutionary, counter-revolutionary or *status quo* theory. It was not the form of theory but its content and its function in society that were at stake.

A critical review of *Explanation in Geography* by Gale was published in 1972. In his reply, Harvey discussed how he then viewed explanation, as such, in geography, having just written 'Revolutionary and Counter-revolutionary Theory in Geography and the Problem of Ghetto Formation'. Before going on, however, two previous statements on theory by Harvey published between 1970 and 1972 will be noted. This will further illustrate the reorientation in Harvey's view of explanation and theory during this period.

When Harvey's first article, 'Locational Change in the Kentish Hop Industry and the Analysis of Land Use Patterns', was reprinted in 1970 in a collection of essays on historical geography, Harvey added a supplementary note which also referred to his doctoral research (Harvey, 1970c). He stated that in 1963 he had 'very little understanding of the importance of location theory in analyses of this kind' (Harvey, 1970c, p. 264). He had found that he had to 'reach out for some kind of adequate framework' to enable him to put his thoughts together. He had therefore constructed a framework out of some 'rudimentary notions of spatial interaction, cumulative causation, and the like'. Harvey concluded that 'a careful integration of theory and empiricism' was central to the progress of historical geography (Harvey, 1970c, p. 265).

In a short article published in *New Movements in the Study and Teaching of Geography* in 1972, but probably written earlier, Harvey summarised the view of theory put forward in *Explanation in Geography*, noting that there had been a 'negative side' to the conscious use of theory construction and model-building techniques. He admitted that 'the models we have built and the theories we have set up over the past decade have not been as general or far-reaching in their coverage as we had hoped' (Harvey, 1972a, p. 29). Such muted comments gave way to more comprehensive criticisms in his reply to Gale's (1972a) review of *Explanation in Geography* (Harvey, 1972f). Here, Harvey put forward a radically different view of theory and explanation from that earlier expressed. The separation between philosophy and methodology in *Explanation in Geography* had led to the separation between the form and the content of theory. This was inadmissible, Harvey believed (Harvey, 1972f, p. 324). Furthermore, the hypothetico-deductive account of theory 'superimposes a formalistic structure over all content and thereby presumes that the world is arranged on the basis of formal principles, which it clearly is not' (Harvey, 1972f, p. 325). Harvey repeated his rejection of *verstehen* and empathy, this time because they could be 'quickly perverted into apology' for the *status quo* (Harvey, 1972f, p. 326). As far as language was concerned, Harvey noted, as in 'Revolutionary and Counter-revolutionary Theory in Geography and the Problem of Ghetto Formation', that positivism used 'fixed stationary categories of thought to deal with a shifting universe'. Categories and meanings should change 'coherently and relationally' to mirror the changing experience which was being portrayed. Harvey noted that

Ollman (1971) had pointed to this as being Marx's approach. Categories were used by Marx 'to mirror the properties of other categories in a whole complex of categories set out in a manner that itself mirrors the real life we seek to portray' (Harvey, 1972f, p. 328). This is what Marx attempted in *Capital*, according to Harvey, and it was a technique which the traditional scholarship of the Anglo-Saxon world, with its logical-empiricist language, was unable to comprehend.

## Marxism and Urban Analysis

Harvey's (1972g) *Society, the City and the Space-economy of Urbanism* and his contribution to *The Housing Market and Code Enforcement in Baltimore* (Harvey *et al.*, 1972) contain versions of parts of Chapters 5 and 6 of *Social Justice and the City* (Harvey, 1973a), although they make no reference to them. It appears that the four pieces were written at about the same time in 1972, although the chapters in *Social Justice and the City* present a more sophisticated account of certain issues than do the other two works.

*Society, the City and the Space-economy of Urbanism* was published in 1972 as a resource paper by the Commission on College Geography of the Association of American Geographers. In it, Harvey focused upon the tension between the established spatial organisation of society and any new form demanded by an emerging social order, given that the city, as a built form, is slow to adapt to change. As such, the paper drew upon three previously published articles (Harvey, 1971, 1972b, 1972c), although a considerable amount of new material was introduced.

Harvey's purpose in the paper was to identify theoretical concepts and orientations appropriate to a revolutionary 'change embracing' perspective that would not simply uphold the *status quo* (Harvey, 1972g, p. 2). In Part 1, he outlined Fried's (1967) distinction between an 'egalitarian' society characterised by 'reciprocity', a 'rank' society characterised by 'redistribution' and a 'stratified' society based upon 'market exchange' (Harvey, 1972g, pp. 3-6). Cities were based upon the extraction of a surplus gained from a surrounding economy and only the rank and stratified societies contained mechanisms to extract and concentrate such a surplus, thus supporting the development of urbanism (Harvey, 1972g, pp. 6-7).

In Part 2 of *Society, the City and the Space-economy of Urbanism*, Harvey analysed the contemporary city as 'a system within which the surplus is generated and appropriated'. He took housing as 'the incision point from whence we can begin to dissect other facets of the urban system' (Harvey, 1972g, p. 15). Harvey outlined the 'use-value' of a house, that is, the set of uses to which a house is put by its user, and its 'exchange-value', terms which, according to Harvey, originated with Adam Smith (1970) and were subsequently detailed by Marx (1967). Next, Harvey looked at a number of case studies, one of which, for instance, considered how the spatial form of a city was affected by changes in the place of residence and the location of job opportunities within a transport network. He extended part of the analysis contained in 'Social Processes, Spatial Form and the Redistribution of Real Income in an Urban System' (Harvey, 1971, pp. 274-7), showing how such changes disadvantaged the poor and redistributed wealth to richer groups. The problem was illustrated with reference to a number of cities in the United States and Harvey concluded, as in a previous article (Harvey, 1972c), that elimination of competitive bidding for use of land would eliminate the resulting inequalities (Harvey, 1972g, pp. 15-25).

Harvey identified the various 'actors' in the housing market, such as the occupiers of housing, landlords, developers, financial institutions and the government, and he noted how they all sought to meet their own, usually economic, interests through the housing market (Harvey, 1972g, pp. 35-46). Here, Harvey used maps and data also included in *The Housing Market and Code Enforcement in Baltimore* (Harvey *et al.*, 1972). He went on to contrast the general sequence of decay and decline in the inner cities with the development of prosperous conditions in the suburbs, pointing out that the controlling factor 'appears to be the policies of financial institutions' (Harvey, 1972g, p. 48).

> The penetration of the market economy into the housing sector, with each actor seeking to realize a profit, naturally gives rise to a situation in which the social surplus is appropriated by some and effectively yielded up by others . . . The market . . . relies upon scarcity in order to function. In order to guarantee the supply of new housing to the rich, it appears that the poor are forced to bear all the burdens of a socially created scarcity . . . The solution to the housing problem in American cities depends upon the willingness to grapple with some fundamental characteristics of the market process. As Engels correctly pointed out, it is difficult to envisage such grappling being able to stop merely at housing (Harvey, 1972g, p. 49).

Harvey concluded by envisaging the possibility of co-operation among people giving rise to a new form of society 'in which the surplus is produced and appropriated for the use of all' (Harvey, 1972g, p. 52).

In July, 1972, Harvey and four others from the Department of Geography and Environmental Engineering at Johns Hopkins University produced a report for the Baltimore Urban Observatory (Harvey *et al.*, 1972). Entitled *The Housing Market and Code Enforcement in Baltimore*, it dealt with the functions of housing code enforcement in relation to the dynamics of the Baltimore City housing market. The authors showed, for instance, how different submarkets were characterised by different rates of return on investment in rental housing and different patterns of investor behaviour. Formal model building was eschewed in the report (Harvey *et al.*, 1972, p. 4.1), although two simple housing inspection models were developed.

It is difficult to isolate Harvey's contribution to this report but it is certain that he wrote most of Chapters 3 and 4 because they contain material in common with an earlier paper (Harvey, 1972g) and with Chapter 5 of *Social Justice and the City*. For instance, the absolute, relative and relational attributes of urban space were outlined, along with the utility of the concepts of exchange value and use value in relation to housing, and a 'blow-out and filter-down' representation of the dynamics of the housing market was presented (Harvey *et al.*, 1972, pp. 3.1-3.3, 3.3-3.5, 4.2-4.7; cf. Harvey, 1973a, pp. 168, 155-7, 172-3). Consequently, any new material in Harvey's contribution to the report will be dealt with in the analysis of Chapter 5 of *Social Justice and the City*.

In 1972, Harvey wrote two papers which were to be published as Chapters 5 and 6 of *Social Justice and the City*. Chapter 5 was entitled 'Use value, exchange value and the theory of urban land use' (Harvey, 1973a, pp. 153-94). In it, Harvey advocated a reconsideration of the concepts and approaches of classical political economy, particularly the Marxist concepts of use value and exchange value. Relying on Ollman (1971, pp. 179-89), Harvey pointed out that Marx used words in a 'relational and dialectical way'.

> Use value and exchange value have no meaning in and of themselves ... For Marx, they take on meaning (come into existence if you will) through their relationship to each other (and to other concepts) and through their relationship to the situations and circumstances under discussion (Harvey, 1973a, p. 154).

Harvey went on to illustrate how Marx brought use value and exchange value into a dialectical relationship with each other through the form they assumed in a commodity (Harvey, 1973a, pp. 155-6). He also noted that Marx viewed a commodity as an expression of a set of social relationships. In capitalist society, for instance, a commodity expresses 'universal alienation' (Harvey, 1973a, p. 156).

Pointing out that marginalist economics separated use value and exchange value, and that geographers, planners and sociologists have dealt with use value only, or have borrowed unquestioningly from marginalist economics, Harvey commented that Marx's approach offered the prospect of 'building a bridge between spatial and economic approaches to urban land-use problems' (Harvey, 1973a, p. 157). He outlined the use value and exchange value of 'land and improvements' and concluded that an adequate theory of urban land-use must be based on 'those catalytic moments . . . when use value and exchange value collide to make commodities' out of land and improvements (Harvey, 1973a, p. 160).

Harvey developed a three-fold view of space. Housing occupied 'absolute space' because it could not be moved around. This conferred monopoly privileges upon the owner.

This concept is not *in itself* an adequate conceptualization of space for formulating urban land-use theory. The distance between points is *relative* because it depends upon the means of transportation, on the perception of distance by actors in the urban scene, and so on . . . We also have to think *relationally* about space for there is an important sense in which a point in space "contains" all other points (this is the case in the analysis of demographic and retail potential for example and it is also crucial for understanding the determination of land value . . .) (Harvey, 1973a, p. 168).

Harvey pointed out that the urban land-use models of Alonso (1964) and Muth (1969) overlooked certain important aspects of the absolute character of space as well as certain aspects of the institutional setting of a capitalist economy. Accordingly, he went on to suggest a 'sequential space-packing model' of the housing market where allocation of housing occurred 'in a sequential manner across an urban space divided into a large but finite number of land parcels' (Harvey, 1973a, p. 168). If the sequential allocation was in order of competitive bidding power, then, given the monopolistic character of house-ownership, 'the rich can command space whereas the poor are trapped in it' (Harvey, 1973a, p. 171).

Microeconomic theories of urban land use, despite their unrealistic assumptions, produced results reasonably consistent with reality. In exploring the reason for this, Harvey surveyed the neo-classical view of rent as a return to a scarce factor of production and contrasted it with Marx's (1967, Volume 3) distinction between monopoly, differential and absolute rent, each of which arise out of different circumstances (Harvey, 1973a, pp. 179-84).

> The power of Marx's analysis of rent lies in the way that he dissects a seemingly homogeneous thing into its component parts and relates those parts to all other aspects of the social structure. Rent is a simple payment to the owners of private property, but it can arise out of a multiplicity of conditions (Harvey, 1973a, p. 184).

The relative success of microeconomic models of urban land-use was due to their exclusive reliance upon the concept of differential rent and to the setting of their analytics in a relative space. Furthermore, their 'assumption of centricity' contributed to their 'appearance of empirical relevance' (Harvey, 1973a, pp. 188-9).

In the concluding section of the chapter, Harvey noted that rent served to allocate land to uses. When use determined value, allocation was 'rational' and led to efficient capitalist production patterns, although as such it was an expensive mechanism. When value determined use, rampant speculation and artificially induced scarcities became part of the allocation process which then operated much less efficiently (Harvey, 1973a, p. 190). Social policy seemed helpless in the face of the latter. 'The rentier will get that pound of flesh no matter what' (Harvey, 1973a, p. 191).

Harvey finally concluded that all land-use theories were specific to a particular mode of production, set of social relations and set of institutions. The 'von Thunen type models' provided a normative theory of land-use based on the concept of rent 'as a shadow price which represents social choices foregone'. Harvey believed that this concept could help to determine socially beneficial land-use decisions consistent with the aims of society and thereby

> provide the basis for revolutionary advances with respect to the creation of socially efficient and humane urban structures ... Theories or models are not in themselves *status quo, revolutionary* or *counter-revolutionary* ... Theories and models only assume one or other of these statuses as they enter into social practice, either through shaping the consciousness of people with respect to the processes which operate around them, or through providing an

analytical framework as a springboard for action (Harvey, 1973a, p. 194).

Chapter 6 of *Social Justice and the City*, entitled 'Urbanism and the city – an interpretive essay' (Harvey, 1973a, pp. 195-284), consists of two parts. In Part 1, Harvey explored the concepts of mode of production, mode of economic integration and social surplus within the context of a general and preliminary enquiry into the 'essential qualities of urbanism'. In Part 2, Harvey went on to examine the relationship between modes of economic integration, the creation of the social surplus and the various historical forms of urbanism. The central thesis of the essay was that

> by bringing together the conceptual frameworks surrounding (1) the surplus concept, (2) the mode of economic integration and (3) concepts of spatial organization, we will arrive at an overall framework for interpreting urbanism and its tangible expression, the city (Harvey, 1973a, pp. 245-6).

When discussing the notion of 'mode of production', Harvey pointed out that Marx (1970) had viewed the relationships of production as the 'real foundation' of society upon which arose a legal and political superstructure. Changes in the economic foundation led sooner or later to the transformation of the whole superstructure. After noting that a mode of production must be defined in relation to a particular social situation, Harvey suggested that 'the mode of production refers to those elements, activities and social relationships which are necessary to produce and reproduce real (material) life' (Harvey, 1973a, p. 199). There were three basic elements to a mode of production, namely, the object of labour or the raw materials existing in nature, the means of labour such as machinery, and labour power. Each society also had a particular mode of economic integration as an integral part of its economic basis, bringing together the various elements of production and diverse socially productive activities (Harvey, 1973a, pp. 199-200). Harvey believed that both Marx and Engels rejected 'simple economic determinism' but viewed superstructural institutions and the state of consciousness in society as 'necessarily both supportive and reflective of conditions in the economic basis of society' (Harvey, 1973a, p. 200). Harvey also wrote that the concept of mode of production was not an 'ideal type' which had conceptual utility but no empirical validity (Harvey, 1973a, p. 203). This implied a rejection of the view of theoretical concepts put forward in his doctoral thesis and in *Explanation in Geography*. There, a theoretical concept was merely an 'instrument' to order a complex reality but Harvey was now viewing theoretical concepts as having empirical content. This reflects Harvey's increasing concern with the 'content' rather than the 'form' of theory.

Harvey went on to distinguish between urbanism as a social form, the city as a built form, and the dominant mode of production. The city and urbanism could function to stabilise a particular mode of production although the city had historically also functioned as a centre of revolution against the established order (Harvey, 1973a, pp. 203-4). Considering the concept of a mode of production as developed by Marx to be too broad and all-embracing to provide the analytical tools to explore the relationship between society and urbanism, Harvey adopted Polanyi's (1968) concept of 'mode of economic integration' and Fried's (1967) concept of 'mode of social organization' which he had outlined in *Society, the City and the Space-economy of Urbanism*. Cities were formed through the geographical concentration of a 'social surplus product' which was produced and concentrated by the mode of economic integration (Harvey, 1973a, p. 216). A social surplus could be taken to be the quantity of material resources over and above subsistence requirements but Harvey noted that 'subsistence' meant something different in different circumstances. 'Each mode of production and each mode of social organization has implicit within it a particular definition of surplus' (Harvey, 1973a, p. 219). Marx went beyond this relativistic view of surplus to define it as the amount of material product set aside to promote improvements in human welfare. In all modes of production so far, except in primitive communism, this surplus could be equated with the product of alienated labour (Harvey, 1973a, pp. 219-20).

Harvey believed that the extraction of surplus labour power did not necessarily give rise to urbanism. Rather, a significant amount of the social surplus had to be concentrated at one point in space. A redistributive mode of economic integration has the ability to concentrate the social surplus but, according to Harvey, it is the market exchange mode that most typically leads to permanent concentrations of surplus value. These concentrations were put into circulation to reap further surplus value, capital accumulation thus taking place (Harvey, 1973a, pp. 226-7). Harvey pointed out that if surplus value was regarded as the manifestation of surplus labour under capitalism, then 'urbanism in capitalist societies can be analysed in terms of the creation, appropriation and circulation of surplus value' (Harvey, 1973a, p. 231). 'Urbanism', Harvey concluded, is a patterning of individual activity which, when aggregated, forms a mode of economic and social integration capable of mobilizing, extracting and concentrating significant quantities of the socially designated surplus product (Harvey, 1973a, p. 239).

In the second part of Chapter 6 of *Social Justice and the City*, Harvey suggested that a revolutionary theory of urbanism need not rewrite previous theories but may redefine the terms contained in them. He investigated how the concepts developed in the first half of the chapter could be used to 'dissect the relationship between urbanism and society in a variety of historical contexts' (Harvey, 1973a, p. 246). There followed four sections, the first on 'patterns in the geographic circulation of the surplus', where Harvey viewed such patterns as 'a moment in a process' so that the history of particular cities could be understood only in terms of 'the circulation of surplus value at a moment in history within a system of cities' (Harvey, 1973a, pp. 249-50). The second section dealt with the cities of medieval Europe within the context of which Harvey examined 'merchant capital' and the relationship between 'town and country' (Harvey, 1973a, pp. 250-61). Section 3 was entitled 'The market exchange process and metropolitan urbanism in the contemporary capitalist world'. Here, Harvey suggested that all aspects of contemporary society were threatened by the potentially destructive power of the market exchange system. Quoting Polanyi (1944), he asserted that 'the alleged commodity "labour power" cannot be shoved about, used indiscriminately, or even left unused, without affecting also the human individual who happens to be the bearer of this peculiar commodity' (Harvey, 1973a, p. 265). Harvey then turned his attention to 'monopoly capitalism' and considered how 'poverty populations' were used to stabilise the capitalist economy by providing an industrial reserve army as well as potential effective demand. Such 'poverty populations' arose out of the institutional creation of scarcity in the commodity of labour power (Harvey, 1973a, pp. 267-73).

The fourth section, also the last section of Chapter 6, was entitled 'Redistribution and reciprocity as countervailing forces to market exchange in the contemporary metropolis'. People resisted viewing themselves as commodities, and relied upon status, prestige and privilege to provide other measures of personal worth. However, this only perpetuated a 'false consciousness', representing issues of market exchange as issues of status and so on, helping to preserve the capitalist *status quo* (Harvey, 1973a, pp. 274-84). Harvey concluded the chapter with a brief assessment of the concepts of modes of economic and social integration.

Even in the contemporary metropolis, with all of its manifest complexity, it appears that these interpretive devices serve us well as we
seek to construct a theory of urbanism which realistically encom-
passes the necessity to concentrate and circulate surplus value as well
as the necessity to construct a space economy in which the various
modes of economic integration can function effectively (Harvey,
1973a, p. 284).

With Chapters 5 and 6 of *Social Justice and the City*, Harvey effectively abandoned the notion that theory should be expressed in a formal
language, given a set of correspondence rules and used to deductively
generate laws. The focus of his work turned to the development of a
conceptual framework within which urbanism and the city could be
analysed. Harvey had, at this stage, accepted Marx's analysis of society
and had gone on to extend it, in Chapter 6, by developing the concepts
of modes of economic and social integration.

In the September 1972 issue of the *Annals of the Association of
American Geographers*, a review by Harvey of Wheatley's (1971) *The
Pivot of the Four Quarters: A Preliminary Enquiry into the Origins and
Character of the Ancient Chinese City* was published (Harvey, 1972h).
Wheatley's book had been described previously by Harvey as 'exquisite'
(Harvey, 1972g, p. 1) and 'brilliant' (Harvey, 1973a, p. 213) although
its stance was that of a 'formless relativism' (Harvey, 1973a, p. 220).
In his review, Harvey repeated his praise of the work and went on to
clarify his disagreement with Wheatley.

At the moment of transition from non-urban to urban forms of
social organization, Wheatley appears to suggest a religious movement, a conceptual change, dominated ... An "effective space" ...
within which a surplus could be extracted for the support of the
central bureaucracy ... [was] created through various political,
social, and economic instruments of religious authority (Harvey,
1972h, p. 510).

Harvey challenged this conclusion by presenting 'a competing hypothesis
which is equally consistent with the facts', a materialist interpretation
based upon Marx's *Grundrisse*. A shift in the economy or mode of production ultimately governed changes in the superstructure of society.
This did not preclude the possibility of ideological influences upon the
economic basis, although these were of 'secondary effect' (Harvey,
1972h, pp. 510-11).

Harvey concluded that he suspected that Wheatley's emphasis upon the religious component of early urbanism stemmed from 'an inferential difficulty which geographers have long encountered, . . . inferring social process from spatial form'. He pointed out that much of Wheatley's evidence had to do with spatial form and it was this which had led him to interpret urban transformation in terms of 'surficial movements in ideology and cosmic symbolism' rather than in terms of movements in the economic basis of society 'which are far harder to discern' (Harvey, 1972h, p. 512).

## Social Justice and the City

*Social Justice and the City* was published in 1973, bringing together four previously published papers (Harvey, 1970a, 1971, 1972b, 1972c), two previously unpublished papers and an introduction and conclusion. The first four papers were 'substantially reproduced' from their originally published versions (Harvey, 1973a, p. 18) with only minor editorial changes. The only exception was at the end of Chapter 4, where Harvey added an edited version of the second half of his reply to comments on the original paper (Harvey, 1972e).

In the introduction to *Social Justice and the City* written in January 1973 (Harvey, 1973a, p. 19), Harvey explained why he had gathered the six papers together. After completing his 'methodological' book, *Explanation in Geography*, he had decided to explore philosophical issues, neglected in that book, in the context of urban planning, urban systems and urbanism. As a result, an 'evolution' was 'provoked' in his views (Harvey, 1973a, p. 9) and the chapters of *Social Justice and the City* represented various points along an 'evolutionary path'. Harvey believed that such an 'evolution'

> seems inevitable if anyone seeks an adequate and appropriate way to bring together a viewpoint established in social and moral philosophy on the one hand and material questions that the condition of the urban centres in the western world point to on the other (Harvey, 1973a, p. 10).

His 'central' and 'overwhelming' concern for several years, Harvey wrote, had been with the relationship between social processes and spatial forms. The book reflected a progression in Harvey's approach to this concern. In the first chapter (Harvey, 1970a), the reconciliation of process and form was analysed as a linguistic problem and a linguistic solution was sought, but in Chapter 5, 'the problem has become one of human practice (in which the linguistic problem is itself embedded) and solutions therefore lie in the realm of human practice' (Harvey, 1973a, p. 10).

Harvey outlined the evolution of his thought in relation to four 'fundamental' and 'interlocking' themes. First, 'the nature of theory' had been viewed initially on the basis of 'an artificial separation of methodology from philosophy'. From this point of view, facts were seen to be separate from values, objects to be independent of subjects and the 'private' process of discovery in science to be separate from the 'public' process of communicating the results (Harvey, 1973a, pp. 11-12). The construction of theory required a manufactured language with arbitrary definitions. Harvey noted that he went on to reject all of these views. Furthermore, verification was viewed initially as a process of establishing the empirical relevance and applicability of abstract propositions but Harvey later came to view theory as needing to be verified through practice. The evolution of his conception of theory, Harvey pointed out, entailed 'a shift away from philosophical idealism towards a materialist interpretation of ideas as they arise in particular historical contexts' (Harvey, 1973a, pp. 12-13).

The second theme through which Harvey illustrated his 'evolution' of thought was that of 'the nature of space'. Initially, the question of 'what is space?' was seen to have a philosophical or linguistic solution independent of everything else. Later, 'space becomes whatever we make of it during the process of analysis . . . Space is neither absolute, relative or relational *in itself*, but it can become one or all simultaneously depending on the circumstances' (Harvey, 1973a, p. 13). Human practice was seen to resolve the proper conceptualisation of space.

The third theme, 'the nature of social justice', was viewed, first, as if social and moral philosophy was a distinct field of enquiry laying down absolute ethical principles which could then be applied to urban questions. This, Harvey admitted, separated fact from value. 'For Marx, the act of observing *is* the act of evaluation' and to separate them was to force a distinction on human practice that has never existed (Harvey, 1973a, p. 15). Harvey regarded the evolution occurring in relation to questions of social justice as that 'from a liberal to a socialist (Marxist) conception of the problem' (Harvey, 1973a, p. 15).

The fourth and last theme Harvey outlined was that of 'the nature of urbanism'. He believed that he had initially treated urbanism as a 'thing in itself', able to be understood as such. But by Chapter 6, he had come to view it as a mirror in which all other aspects of society were reflected (Harvey, 1973a, pp. 16-17). Harvey went on to note that the emergence of Marx's analysis as a guide to enquiry (by which token I suppose I am likely to be categorized as a "Marxist" of sorts) requires some further comment. I do not turn to it out of some *a priori* sense of its inherent superiority (although I find myself naturally in tune with its general presupposition of and commitment to change), but because I can find no other way of accomplishing what I set out to do or of understanding what has to be understood (Harvey, 1973a, p. 17).

Marx had regarded 'ideology' as the 'unaware' expression of underlying ideas and beliefs attaching to a particular social situation. In the West, Harvey pointed out, 'ideology' was the 'aware' and critical exposition of ideas in a social context. He concluded that Chapters 4, 5 and 6 of *Social Justice and the City* were ideological in the Western sense whereas the first three chapters were ideological in the Marxist sense (Harvey, 1973a, p. 18). Harvey consequently divided the book into three parts. Part 1, the first three chapters, he entitled 'Liberal Formulations', Part 2 (Chapters 4, 5 and 6) was designated 'Socialist Formulations' and Part 3, consisting of Chapter 7 (the conclusion), was given the title of 'Synthesis'. The title of the third part implied that Part 1 was the 'thesis' and Part 2 the 'antithesis', and thus the structure of the book reflected Marx's dialectical method. The synthesis, Chapter 7, will next be given attention.

### Marx's Method

In Chapter 7, the conclusion, of *Social Justice and the City*, Harvey developed in more detail his view of Marx's method, and assessed its contribution towards an understanding of urban phenomena. Relying upon Ollman (1971, 1972), Harvey presented Marx's method in terms of its 'ontology' and its 'epistemology'.

'An ontology is a theory of what exists. To say, therefore, that something has ontological status is to say that it exists' (Harvey, 1973a, p. 288). Quoting Ollman (1972, p. 8), Harvey believed that

the twin pillars of Marx's ontology are his conception of reality as a totality of internally related parts, and his conception of these parts as expandable relations such that each one in its fullness can represent the totality (Harvey, 1973a, p. 288).

The wood engraving reproduced on the cover of *Social Justice and the City*, Escher's 'Doric Columns', shows two very similar yet subtly different stone pillars, one upside down alongside the other. The top of each pillar is attached to the base of the other, seeming almost to be flowing out of it. The two columns may be viewed as a graphic illustration of 'twin pillars' consisting of a 'totality of internally related parts', each column in its 'fullness' representing the 'totality'.

Harvey pointed out that a 'totality' might be viewed as an aggregate of elements or as an 'emergent' phenomenon, existing independently of its parts. But Marx's view, according to Harvey, was an 'operational structuralist' one, which has been illustrated in Figure 3.1. Harvey cited Piaget's (1970, p. 9) definition of operational structuralism where 'the relations among elements' are of primary importance and 'the whole . . . is consequent on the system's laws of composition, or the elements' (Harvey, 1973a, p. 288). The relationships between elements within the totality were regarded as expressing certain 'transformation rules' through which the totality was transformed. The totality was structured by the 'elaboration' of the relationships within it and it, in turn, shaped the parts so that each part functioned to preserve the whole.

> Capitalism, for example, seeks to shape the elements and relationships within itself in such a way that capitalism is reproduced as an ongoing system. Consequently, we can interpret the relationships within the totality according to the way in which they function to preserve and reproduce it (Harvey, 1973a, p. 289).

Furthermore, each element reflected the characteristics of the totality. Concepts such as 'labour power' and 'surplus', for instance, have to be treated as reflections of all the social relationships of a particular mode of production (Harvey, 1973a, p. 289).

But relationships occurring in society are not always harmonious. Contradictions arise, leading to conflict. A transformation of the totality takes place through the resolution of these conflicts, the totality is restructured and, in turn, alters the 'definition, meaning and function of the elements and relationships within the whole'. Then new conflicts and contradictions emerge to replace old ones (Harvey, 1973a, p. 289). Harvey pointed out that Marx's ontology implied that research ought to be directed to discovering the transformation rules by which society was being restructured, not to seeking out 'causes', which presuppose 'atomistic association', or 'descriptive laws' governing the evolution of totalities independent of their parts (Harvey, 1973a, pp. 289-90). A totality consisted of a number of separate 'structures', according to Harvey. A 'structure' was not a 'thing' or an 'action' and it could not be established through observation. Rather, as with a totality, a structure was 'a system of internal relations which is in the process of being structured through its own transformation rules' (Harvey, 1973a, p. 290).

Figure 3.1: Marx's Operational Structuralist View of Society (based upon Harvey, 1973a, 288-92).

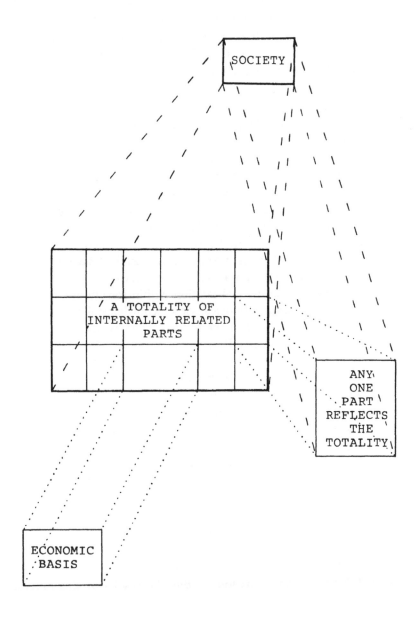

Marx asserted the primacy of the economic basis. In doing so, he was suggesting, according to Harvey, that the relationships between structures were themselves structured in some way within the totality, such that

> in a conflict between the evolution of the economic basis of society and the elements in the superstructure, it is the latter that have to give way, adapt, or be eliminated. Some structures are therefore regarded as more basic than others within a totality . . . When we attempt to view society as a totality, then ultimately everything has to be related to the structures in the economic basis of society (Harvey, 1973a, p. 292).

In other words, for Marx, the production and reproduction of material existence was the 'starting point and end point' for discerning the relationships between structures within the totality (Harvey, 1973a, p. 292). Later, consistent with Marx's viewpoint, Harvey stated that he regarded 'the channels through which surplus value circulates as the arteries through which course all of the relationships and interactions which define the totality of society' (Harvey, 1973a, p. 312).

'Epistemology', according to Harvey, 'seeks to uncover the procedures and conditions that make knowledge possible' (Harvey, 1973a, p. 296). For Marx, knowledge was part of human experience and practice and was viewed as an internal relation within society as a totality. As such, it implied Marx's ontological view. Knowledge is a productive activity and may thus be part of the process whereby society is transformed (Harvey, 1973a, p. 296). Relying on Piaget (1972a), Harvey outlined the view of 'traditional empiricism' that all cognitive information had its source in objects so that the knowing subject was 'instructed' by what was 'outside' of him or her. 'A *priorism* and innatism', on the other hand, assumed that the subject 'possesses from the start endogenous structures' which were imposed upon objects (Harvey, 1973a, p. 297). Marx's epistemology was what Piaget (1970) called 'constructivism'. Here, the subject was regarded as both structuring and being structured by the object (Figure 3.2). The subject assimilated and transformed perceptions and images into concepts through 'reflective abstraction'. These concepts did not have an independent existence and were not universal abstractions true for all time.

> The structure of knowledge can be transformed, it is true, by its own internal laws of transformation . . . But the results of this process

have to be interpreted in terms of the relationships they express within the totality of which they are a part. Concepts are "produced" under certain conditions (including a pre-existing set of concepts) while they also have to be seen as producing agents in a social situation (Harvey, 1973a, p. 298).

Figure 3.2: Marx's Constructivist Epistemology Compared With Other Epistemologies (based upon Harvey, 1973a, 297-8; cf. Gregory, 1978, 58, fig. 2).

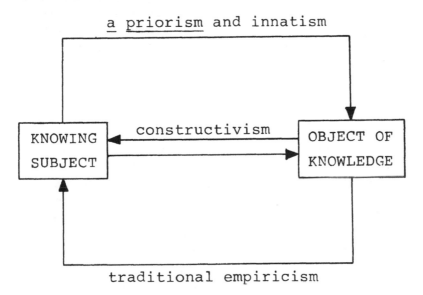

According to Harvey, this meant that it was irrelevant to ask whether concepts were true or false. It was rather what produced them and what they served to produce that was important. Hence arose the distinction between revolutionary, counter-revolutionary and *status quo* theories (Harvey, 1973a, p. 298). Thus Harvey concluded that in so far as knowledge became a material force in society, its internal restructuring, due to the resolution of internal contradictions, could exert influence throughout the totality of society and ultimately be 'registered in the economic basis' (Harvey, 1973a, p. 299).

Harvey noted that the evolution of his thought as expressed in the chapters of *Social Justice and the City* brought about 'a convergence towards an ontological and epistemological position akin to that held by Marx' (Harvey, 1973a, p. 301). Such an evolution was a 'necessity' if the dilemmas that beset the early chapters of the book were to be resolved. These dilemmas arose out of a consideration of 'pressing and serious problems' in the world in the late 1960s. Such problems demanded an interdisciplinary approach which could only be provided by 'the operational structuralist method which Marx practises and which Ollman and Piaget describe' (Harvey, 1973a, p. 302).

Having established his conclusions regarding the character of Marx's method, Harvey went on, in the second part of Chapter 7 of *Social Justice and the City*, to examine how that method had yielded insights into the nature of urbanism. He outlined two possible approaches to urbanism using Marx's method, the first viewing urbanism as a separate structure with its own laws of structuring and inner transformation, the second viewing urbanism as the expression of a set of relationships embedded in some broader structure such as the social relations of production (Harvey, 1973a, pp. 303-4). Briefly surveying the history of urbanism, Harvey pointed out that urbanism first emerged as a separate structure with a certain amount of autonomy despite being part of the superstructure of society (Harvey, 1973a, pp. 304-5). He viewed modern urbanism as a 'self-sustaining entity which expresses and fashions relationships with other structures in the totality' (Harvey, 1973a, p. 307). In relation to contemporary society, Lefebvre's main thesis was that industrialisation was being produced by urbanism whereas urbanism was once produced by industrialisation (Lefebvre, 1970, 1972). Lefebvre had arrived at this thesis through 'construction by negation'. But Harvey, while agreeing with Lefebvre on most issues, suggested that this 'device' provided an hypothesis but not a proof. Harvey also believed that the hypothesis could not be substantiated 'at this point in history' (Harvey, 1973a, pp. 305-13). He concluded:

> A genuinely humanizing urbanism has yet to be brought into being. It remains for revolutionary theory to chart the path from an urbanism based in exploitation to an urbanism appropriate for the human species. And it remains for revolutionary practice to accomplish such a transformation (Harvey, 1973a, p. 314).

In June, 1973, Harvey delivered the annual Norma Wilkinson memorial lecture at the University of Reading, England (Harvey, 1973b). Entitled *A Question of Method for a Matter of Survival*, it was a case study of the influence of methodology upon 'scientific' research. It contained material in common with Chapter 7 of *Social Justice and the City* but extended the account there in order to compare Marx's method with that of Malthus and Ricardo in the context of what Harvey called 'the population-resources debate'.

Harvey identified Malthus's method as 'logical empiricism', an approach which emphasised observed evidence and *a priori* deduction. Such a method, Harvey asserted, produced a *status quo* theory (Harvey, 1973b, pp. 1-2). This method had become known as 'the scientific method' and was viewed as beyond criticism because science had been defined as being ideology-free. Such a definition 'is a normative statement of aims and cannot be justified by science's own methods', Harvey argued. 'The definition . . . originates from a source external to scientific enquiry' and was thus not ideology-free (Harvey, 1973b, p. 3). Harvey went on to examine Malthus's theories, illustrating their ideological character in support of the landed aristocracy.

Ricardo's method was that of 'normative deductive analytics', that is, Ricardo appealed to *a priori* deductive reasoning, and he constructed

> a carefully fashioned analytic theory which had little need for an empirical base . . . He took the structure of competitive capitalism and through the power of abstraction created a system of thought which embodied the aspirations and behavior of an idealized "economic man" (Harvey, 1973b, p. 10).

This method allowed Ricardo to attack the *status quo* and to support the aspirations of the rising bourgeoisie.

Harvey outlined Marx's criticism of the views of both Malthus and Ricardo and noted how the differences in their views could be traced back to their methods. 'Each of these methods generates quite different conclusions' (Harvey, 1973b, p. 16). In a section entitled 'The political implications of population theory', Harvey argued that concepts have 'ideological content'. Using 'subsistence', 'resource' and 'scarcity' as examples, he showed how the solution to the problem of 'overpopulation' in relation to the world's resources was dependent upon the ideological viewpoint taken. Noting the need to 're-constitute our notions of method' in discussing the population-resources debate, Harvey went on to look at the view of Marx's method which he had presented in the first part of Chapter 7 of *Social Justice and the City*. He pointed out that he was concerned with 'Marx's method', not the 'Marxist method' associated with a 'somewhat misguided' Engels (1946) and Lenin (1927) (Harvey, 1973b, p. 27). In outlining Marx's operational structuralism (Harvey, 1973b, pp. 28-39), Harvey noted that this directed research towards 'discovering the transformation rules and processes of self-regulation whereby society is constantly being restructured' (Harvey, 1973b, p. 30). In Chapter 7 of *Social Justice and the City* (Harvey, 1973a, p. 289), Harvey had only referred to 'transformation rules', omitting 'processes of self-regulation', which considerably narrowed the scope of potential research.

Harvey went on to suggest that the results of logical empiricism and deductive analytics were not irrelevant or wrong. Rather, they required 'enrichment' through the integrating power of 'higher order' concepts. 'And it is exactly this kind of enrichment that Marx's method can provide in the population-resources debate' (Harvey, 1973b, p. 39). Furthermore, Marx's method was the only one capable of dealing with the complexities of the population-resources debate. These conclusions, Harvey noted, would be 'unpalatable' to most researchers because they sounded so ideological. But failure to utilise Marx's method was

> to court ignorance on a matter as serious as the survival of the human species. And if ignorance is the result of the ideological belief that science is and ought to be ideology free, then it is our hidden ideology that is the most serious barrier to enquiry. And if, out of ignorance, we participate in the politics of repression and the politics of fear, we are doing so largely as a consequence of the ideological claim to be ideology free. But then, perhaps, it was precisely that participation that the claim to be ideology free was designed to elicit all along (Harvey, 1973b, p. 41).

In 1974, two heavily edited and considerably shortened versions of *A Question of Method for a Matter of Survival* were published as articles. One was entitled 'Ideology and Population Theory' and appeared in the *International Journal of Health Services* (Harvey, 1974c) while the second one, 'Population, Resources, and the Ideology of Science', was published in *Economic Geography* (Harvey, 1974d). There was very little difference between the two articles, in which Harvey attacked the ethical and ideological neutrality of science. He outlined the methods of Malthus, Ricardo and Marx in relation to the population-resources debate, coming to the same conclusions as in *A Question of Method for a Matter of Survival*. There are perhaps two differences of note between the memorial lecture and the two journal articles. In the latter, Harvey characterised Ricardo's method not as 'normative deductive analytics' but as an 'analytic model-building' approach (Harvey, 1974c, p. 522, 1974d, p. 262), which made its relationship to modern geography much clearer. Harvey also did not use the term 'operational structuralism' to describe Marx's method but rather called it 'dialectical materialism' (Harvey, 1974c, p. 524, 1974d, p. 265), relying only upon Ollman's (1971, 1973) account of Marx's ontology as a conception of reality 'as a totality of internally related parts', omitting any reference here to Piaget. When Harvey dealt with Marx's epistemology, he used Piaget (1970) as his source for its designation as 'constructivist' as in *A Question of Method for a Matter of Survival* (Harvey, 1974c, p. 527, 1974d, p. 267).

In his four works on Marx's method, Harvey clearly outlined what he saw to be the ontological and epistemological foundations of Marx's method. In Chapter 7 of *Social Justice and the City* he also suggested how that method could be used to gain insight into the nature of urbanism. In three other publications (Harvey, 1973b, 1974c, 1974d), he used the population-resources debate to contrast Marx's method with that of Malthus and Ricardo, whose approaches were presented as thinly disguised representatives of the logical empiricist and model-building approaches in modern Anglo-American positivist geography.

In January 1974, Harvey delivered a paper entitled 'What Kind of Geography for what Kind of Public Policy?' at the annual conference of the Institute of British Geographers in Norwich, England (Harvey, 1974a). Pointing out how Pinochet, a geographer by training, had actively transformed the geography of Chile's health care system in order to serve the rich to the detriment of the poor, Harvey invited geographers to reconsider both the kind of geography they were developing and their relationship to the corporate state. He placed the evolution of geography as a discipline against a background of 'changing social necessities' (Harvey, 1974a, p. 19). In Britain, around 1945, geography had become reoriented in its focus. Previously concerned with 'the technics and mechanics of the management of Empire', geography then sought to contribute to 'the technics and mechanics of urban, regional and environmental management' (Harvey, 1974a, p. 20). And Harvey identified the rise of the corporate state as providing the external conditions to which post-World War II geography adapted.

The tension felt by geographers between humanism, with its associated sense of moral obligation, and serving the needs of the corporate state had in part been resolved by separating facts from values, according to Harvey (1974a, p. 22). But this dualism was artificial and led to support of the *status quo*. Taking his cue from Marx, Harvey advocated that geographers should attempt to subvert the ethos of the corporate state from within by collapsing the dualisms inherent in *status quo* geography. He viewed this as a step towards the creation of an 'incorporated state' which reflected 'the creative needs of people struggling to control the social conditions of their own existence in an essentially human way' (Harvey, 1974a, p. 23).

In this paper, Harvey invited geographers to adopt a Marxist perspective upon their work. He outlined the role that a geography based upon Marx's method should play in a capitalist society. Having now fully embraced Marx's method, Harvey was to go on to apply it in an extended study of urbanisation in the United States. This research programme is the subject of Chapter 4.

## From Logical Empiricist to Marxist Geography

The 'internal history' of Harvey's rejection of logical empiricism and adoption of a Marxist perspective may be summarised as follows. Harvey became aware that avoiding the task of defining a coherent social objective nevertheless implied a set of social, political and ethical judgements, namely, that the current conditions extrapolated into the future were acceptable (Harvey, 1971, pp. 267-8, 1972b, p. 118). At the same time, he revealed a dissatisfaction with the injustices and inequalities characterising the current conditions, or *status quo*, of capitalist society (Harvey, 1970a, p. 269, 1971, p. 100). This led him to the view that there could be no such thing as a neutral, objective and value-free theory. Theories either upheld the *status quo*, as did those based on logical empiricism, or held out the possibility of humanising social change, as did those based on Marxism (Harvey, 1972c, pp. 129-30). Harvey chose Marxism.

From the point of view of an 'internal history', if Harvey had decided to continue to examine theories in an abstract manner, as he had in *Explanation in Geography*, it is unlikely that he would have adopted a Marxist approach so quickly, if at all. It was because he decided to explore the incorporation of social justice in a planning context (what might be referred to as a 'material' context) that he concluded that objective, value-free theories effectively upheld the economic, social and political *status quo*. And this conclusion ultimately led him to adopt a Marxist viewpoint. In the introduction to *Social Justice and the City*, Harvey stated that he did not regard the history of his reorientation in approach as idiosyncratic to himself.

It is the sort of history that seems inevitable if anyone seeks an adequate and appropriate way to bring together a viewpoint established

in social and moral philosophy on the one hand and material questions that the condition of the urban centres in the western world
point to on the other (Harvey, 1973a, p. 10).

Although many socially and politically active non-Marxist geographers
would disagree with Harvey on this point, it could be argued that the
examination of the effects of applying logical empiricist theory in a
planning context would undermine belief in logical empiricism's social
and political objectivity and neutrality. However, even then, Marxism is
not the only alternative to a straightforward logical empiricist approach,
as Olsson (1972) and Getis (1972) pointed out.

In Chapters 5 and 6 of *Social Justice and the City*, Harvey developed
a conceptual framework based upon Marx's philosophy, a framework
from which he could criticise positivist theories regarding the city as
well as build up an alternative theoretical approach. In doing so, he
relied upon, first, Ollman's (1971, 1973) view that Marx's work assumed
a 'philosophy of internal relations' and, secondly, the view that Marx's
method combined what Piaget (1970, 1972a) described as an 'operational structuralist' ontology and a 'constructivist' epistemology. To
interpret Marx in this way is to take account of Marx's earlier, less
deterministic, writings in order to understand his later works. Marx's
early works were not available in English until decades after his later
ones. His doctoral thesis, *Critique of Hegel's Philosophy of the State*,
*Economic and Philosophic Manuscripts of 1844* and *The German
Ideology*, all written before the end of 1846, were never published by
him. In 1927, a complete German edition of Marx's early writings was
produced by the Marx-Engels Institute in Moscow but it was not until
much later that English translations became available. In 1961, for
instance, the *Economic and Philosophic Manuscripts of 1844* were
published in English for the first time (McLellan, 1972, pp. 268-73).
Marx's later writings, especially *Capital*, which was first published in
German in 1867, were translated into English much earlier. For instance,
an edited English version of *Capital* appeared in 1932 (McLellan, 1971,
p. 229). Without access to his earlier writings, it was much easier to read
Marx as a mechanistic determinist. With access to his earlier writings,
which reflected the influence of Hegel's philosophy, it was possible to
view Marx as a humanist and as much less deterministic in approach
(Ollman, 1971, pp. 3-11). Ollman presented an account of Marx's
method in the light of his earlier writings and of 'the philosophical
tradition in which he was nourished' (Ollman, 1971, p. 29). Ollman
believed that Marx followed Hegel, Leibniz and Spinoza in seeking 'the

meaning of things and/or of the terms which characterise them in their relations inside the whole' (Ollman, 1971, p. 29). Piaget developed a view of 'operational structuralism' and 'constructivism' in a context completely independent of Marx. Harvey himself made the connection between Ollman's account of Marx's method and Piaget's view of operational structuralism. Where Piaget had noted his convergence with Marx (Piaget, 1972b, p. 204), it was primarily with reference to the dialectic, not to operational structuralism as implied by Harvey (1973a, p. 287ff). Harvey thus made the connection between Marx's epistemology and Piaget's (1972a, p. 19) account of constructivism without reference to any other source. Piaget, for instance, did not note the relationship.

In Chapter 7 of *Social Justice and the City* and his reply to Gale's review of *Explanation in Geography* (Harvey, 1972f), Harvey rejected the separation between philosophy and methodology which was fundamental to *Explanation in Geography*. Having viewed philosophy as studying scientists' individually held values, Harvey came to approach scholarship as a whole within a social and communal context. He further came to view methodology as based upon primarily ontological but also epistemological assumptions, whereas he had previously believed that methodology dealt with the logic of explanation quite separate from philosophical considerations. Methodology had come to have a broader meaning to Harvey. It also had a greater influence upon research than he had earlier believed. In relation to the population-resources debate, for instance, Harvey was able to demonstrate that a certain methodology could 'generate' certain conclusions (Harvey, 1973b, 1974c, 1974d).

Harvey's main criticism of logical empiricism has been noted earlier, that is, that logical empiricist theories, when used in urban or territorial planning or in devising solutions to the population-resources problem, simply extrapolated all the defects of the *status quo* into the future. This stemmed from the separation of fact and value in logical empiricism, that is, from the refusal to incorporate value-laden objectives such as social justice into theory and from restricting analyses to a factual, yet defective, reality. Harvey also criticised logical empiricism for emphasising 'form' to the exclusion of 'content'. It was more concerned with the hypothetico-deductive structure of theory and the verification of abstract propositions than with understanding the complexity of the city and bringing about humanising change.

Harvey developed a three-fold classification of theory, namely, revolutionary, counter-revolutionary and *status quo* theory. Here, a theory was classified in terms of how it entered into 'human practice', for instance, in terms of the role it potentially played in urban planning. Neutrality and objectivity were thus defined in relation to the potential effect of planning upon society. Previously, in *Explanation in Geography*, a theory had been classified in terms of formal criteria, that is, the extent to which it was expressed in a theoretical language as a hypothetico-deductive system (Harvey, 1969a, pp. 97-9). This change in Harvey's approach to the classification of theory reflected the change of emphasis in his work from 'form' to 'content' and from 'idealism' to 'materialism' (Harvey, 1973a, pp. 12-13).

Logical empiricism arbitrarily assigned fixed definitions to terms when, according to Harvey, reality is constantly changing. Marx's 'relational' approach, on the other hand, allowed concepts to change as reality changed. Different modes of production and modes of social organisation, for instance, implied different definitions of 'surplus' (Harvey, 1973a, p. 219). Whereas logical empiricism tended to take an atomistic view of society, Marx's method enabled the researcher to view each facet of society as related to and reflecting every other. In conclusion, for Harvey, Marx's method was more powerful, more realistic, more practical and more critical than that of logical empiricism. As Harvey himself put it, 'I can find no other way of accomplishing what I set out to do or of understanding what has to be understood' (Harvey, 1973a, p. 17).

Harvey was critical of Marx's views on a number of occasions. In Chapter 6 of *Social Justice and the City*, for instance, Harvey suggested that the mode of production concept was 'too broad and all-embracing' to use as a tool to 'dissect' the relationship between urbanism and society. He went on to use Polanyi's (1968) concept of 'mode of economic integration' and Fried's (1967) concept of 'mode of social organization' (Harvey, 1973a, p. 206). Elsewhere, Harvey indicated that Marx had underestimated the spatial component of the economy, particularly the way in which distance could allow owners of land and property to gain absolute and monopoly rents (Harvey, 1973a, p. 183). The influence of Marx's thought as a 'guide to enquiry' (Harvey, 1973a, p. 17), however, is overwhelming compared with these details. In the first example noted above, modes of economic integration and social organisation were used by Harvey as analytical tools within a framework based upon Marx's view that society is made up of substructural and superstructural elements related to and reflecting each other.

In his comments on Harvey's 'Revolutionary and Counter-revolutionary Theory in Geography and the Problem of Ghetto Formation' (Harvey, 1972d), Folke had noted that theory ought not be developed first and then put into practice. Rather, theory and practice should be dialectically integrated (Folke, 1972, p. 17). Harvey believed that theory *was* practice (Harvey, 1973a, p. 12) but he also held the view that theory could be developed quite apart from practice. Thus, 'particular theories or models are not in themselves *status quo, revolutionary* or *counter-revolutionary*' but they 'only assume one or other of these statuses as they enter into social practice' (Harvey, 1973a, p. 194). Furthermore, revolutionary theory was to 'chart the path' for revolutionary practice (Harvey, 1973a, p. 314), implying that theory was to be developed first and then practice could proceed, the very view that Folke had warned against.

Harvey believed that a theory could become either a revolutionary, counter-revolutionary or *status quo* theory depending upon how it entered into human practice. A positivist theory, which accurately described reality and was used to bring about revolutionary material change, could thus be designated as revolutionary (Harvey, 1972e, p. 41, 1973a, p. 194, 1973c, p. 87). But Harvey also called for the reintroduction of the terminology of classical political economy, including such terms as use value and exchange value, and noted the need to develop a conceptual framework based upon the concepts of surplus and modes of economic integration and social organisation in order to provide an overall framework for interpreting urbanism and the city (Harvey, 1973a, Chapters 5 and 6). Elsewhere, Harvey illustrated that a certain method generated a particular set of conclusions (Harvey, 1973b, 1974c, 1974d). From this point of view, Harvey implied that a revolutionary theory ought to be constructed using terms and conceptual frameworks based upon Marx's view of society and utilising Marx's method, analysing a defective reality from the viewpoint of the exploited, incorporating the need for revolutionary change. But if any theory could become a revolutionary (or a counter-revolutionary) theory, depending upon the way in which it entered human practice, why was it necessary to develop Marxist *theory*? Would not the development of Marxist *practice* be much more fruitful? Apparently Harvey believed that theory was crucial to practice for, in the mid-1970s, he went on to develop a Marxist *theory* of urbanisation under advanced capitalism.

In his reply to Gale's critical review of *Explanation in Geography*, Harvey explicitly rejected the hypothetico-deductive view of theory (Harvey, 1972f, p. 324). It imposed formal solutions upon informal problems. A theoretical language ought to be similar to a natural language, having a 'looseness of structure and an inherent ambiguity which permits us to capture part of that shifting movement [of reality] while using the same words' (Harvey, 1972f, p. 327). Thus Harvey rejected the logical empiricist emphasis on formalisation. However, in accepting Marxist theory as a 'guide to enquiry' (Harvey, 1973a, p. 17), Harvey effectively treated it as a source of fruitful hypotheses, that is, as an *a priori* theory from which testable hypotheses could be deduced. To this extent, Harvey utilised Marxist theory in a hypothetico-deductive manner.

After 1969, Harvey nowhere explicitly discussed Hempel's deductive-nomological model of scientific explanation. His outline of Marx's approach included the recognition of dialectical logic (Harvey, 1972c, p. 11), but in the studies in Chapters 5 and 6 of *Social Justice and the City*, for instance, he did not make use of it. Instead, he deduced certain ideas from Marx for which empirical support was then sought. As far as laws were concerned, Harvey referred to the 'inner laws of trans-formation' of society and of knowledge that were an integral part of Marx's operational structuralism (Harvey, 1973a, pp. 296, 298). He also noted that research ought to be directed towards discovering these laws or 'rules' as well as those 'processes' structuring society in the first place (Harvey, 1973a, p. 289, 1973b, p. 30). It is not clear how Harvey en-visaged that a researcher would discover these laws, rules and processes nor, once obtained, how the researcher would use them in explanation. The question of Harvey's view of the validity of the hypothetico-deductive structure of theory and the deductive-nomological model of explanation will be examined again when his application of Marx's method to certain issues in urban geography is analysed in Chapter 4.

At the beginning of the 1970s, Harvey advocated that social science research should be interdisciplinary. In his 1970 paper, for instance, he explored the relationship between geography and sociology in urban studies (Harvey, 1970a). By 1972, Harvey was suggesting that all dis-ciplinary boundaries were counter-revolutionary, allowing the State to 'divide and rule' the application of knowledge. Reality had to be approached 'in non- or meta-disciplinary terms' because 'inter-, multi- and cross-disciplinary studies' never succeeded. But Harvey believed that the first step towards overcoming disciplinary boundaries entailed 'within-disciplinary questioning and re-formulation of theory' in geo-graphy (Harvey, 1972e, p. 40). He was eventually to conclude that the only method capable of creating an interdisciplinary theory of urbanism and of dealing with the complexities of a population-resources system in an integrative fashion was Marx's operational structuralism (Harvey, 1973a, p. 302, 1973b, p. 41, 1974c, p. 536, 1974d, p. 276).

Harvey has admitted that Marx's method took a materialist approach which gave due weight to objective social conditions (Harvey, 1972c, pp. 10-11). Marx's materialism entailed the view that the economic basis was primary in structuring relationships within society. But Harvey took pains to point out that this did not preclude the influence upon the substructure of superstructural elements, which included theory (Harvey, 1973a, pp. 298-300). This could only occur, however, in so far as theory became a 'material force' in society (Harvey, 1973a, p. 298). For Harvey, then, a materialist approach did not mean that theory could not effect change in society. Rather, theory had to become part of 'human practice' in order to do so. Thus Harvey attempted to affirm a materialist perspective without adopting a simplistic view of society as consisting of a passive superstructure mirroring an active economic substructure. This materialism also functioned as an interpretive device. In his review of Wheatley's (1971) book, for instance, Harvey challenged the view that 'a religious movement, a conceptual change' dominated the transition from non-urban to urban forms of social organisation (Harvey, 1972h, p. 510). Harvey inverted this view and offered an alternative hypothesis, that a shift in the economy or mode of production gave rise to changes in the superstructure of society (Harvey, 1972h, p. 511).

At times, Harvey's materialism led him towards economic reductionism and determinism. He regarded 'the channels through which surplus value circulates' as 'the arteries through which course all of the relationships' of society (Harvey, 1973a, p. 312) and his analysis of urbanism in capitalist societies in Chapter 6 of *Social Justice and the City* centred on 'the creation, appropriation and circulation of surplus value' (Harvey, 1973a, p. 231). According to Harvey, the economic aspect of society was viewed by Marx as the 'basis', that is, as 'the starting point and end point for tracing out the relationships between structures within the totality' (Harvey, 1973a, p. 292). Harvey thus stressed the economic aspect as the key to the analysis of society. The economic substructure ultimately determined the form of the ideological superstructure (Harvey, 1973a, p. 200). However, Harvey also put forward the viewpoint that neither Marx nor Engels viewed 'the economic element' as the '*only* determining one' (Harvey, 1973a, p. 198). Marx's assertion of the 'primacy of the economic basis' was only to 'suggest' that in a conflict between substructural and superstructural elements, the substructural element would ultimately prevail. Elsewhere, Harvey put forward the viewpoint that people were not simply economic beings. The human individual was a physical, psychological and moral entity as well as the bearer of 'labour power' (Harvey, 1973a, p. 265). There was thus some ambiguity regarding the way in which Harvey viewed the economic aspect. He did not wish to simply reduce reality to its economic functioning but he wished to make use of the powerful analysis of Marx's method which implied the primacy of an economic basis.

Harvey was aware that some viewed Marx's method as materialistic, reductionistic and deterministic, having once held that view himself (Harvey, 1969a, pp. 425-7). He noted how both 'Russia after Lenin's death' and 'Western sociology' had effectively presented Marx's thought without conveying its humanistic and revolutionary essence (Harvey, 1972c, pp. 8-9). He carefully distinguished between 'Marx's method' and 'Marxist method', describing the latter as 'misguided' (Harvey, 1973b, p. 27). These observations and distinctions were an attempt by Harvey to present Marx's method as a materialist approach which was neither simply reductionistic nor deterministic but which gave an important role to revolutionary theory in bringing about revolutionary change.

Gouldner (1980, pp. 32-63) has argued that there are two Marxisms, namely, critical and scientific Marxism, both having their origins in the tension between freedom and necessity within Marx's own writings (cf. Van der Hoeven, 1976, p. 83). Scientific Marxism, as an 'ideal type', emphasises, for instance, the lawful regularities inherent in society, the discontinuity between Marx and Hegel and a structuralist view of society in which impersonal structures are the true actors. Critical Marxism, on the other hand, emphasises the importance of human decisions in bringing about change, the continuity between Marx and Hegel and society's unique and different character at different stages in its own development as an organic totality within which people act (Gouldner, 1980, pp. 58-60). Gouldner believed that these two long-established trends in Marxist thought are both integral parts of Marxism, presupposing at the same time as opposing each other (Gouldner, 1980, pp. 34, 54; cf. Dooyeweerd, 1953, p. 210, 1960, pp. 45-51). He also believed that the writings of a specific person could not simply be reduced to one of these ideal types. Harvey's struggle to avoid reductionism and determinism, and yet insist upon viewing the economic aspect of society as the substructure of a superstructure which ultimately reflected it, is a struggle with the tension between freedom and necessity in Marx's writings which Marxists inherit. Harvey's motives and aims seemed to emphasise the freedom pole of the tension, whereas his analysis of the structure of society tended to emphasise the necessity pole. This is the fundamental ambiguity in Harvey's writings over the

1972-1974 period. In Chapter 4, an opportunity will be taken to explore the way in which the freedom/necessity tension was reflected in Harvey's later research in which he utilised Marx's method in urban studies.

## Note

1. I am grateful to Professor C. Duncan for the opportunity to have participated in his course in contemporary geographical thought at the University of Waikato. The following section on the philosophy of geography in the first half of the 1970s arises to a large extent out of my experience of that course.

# APPLYING MARXIST METHOD: URBANISM UNDER CAPITALISM AND THE MARXIST ANALYTICAL FRAMEWORK

During the late 1970s and into the 1980s, Harvey conducted research into the post-World War II urbanisation process in advanced capitalist countries. He was assisted by colleagues and students from the Department of Geography and Environmental Engineering at Johns Hopkins University, and he acknowledged in particular the aid and advice of Chatterjee, Klugman and Walker. This programme of research had its beginnings in 1974, when Harvey sought to apply Marx's operational structuralist method to an analysis of the Baltimore housing market. Throughout the second half of the decade, Harvey focused his attention upon the economic, social and political setting of the Baltimore housing market within the context of the United States as an advanced capitalist nation. The theoretical framework he eventually constructed was based upon the twin themes of the accumulation of capital and class struggle. Harvey acknowledged that many of his articles were drawn from a forthcoming book on the urban process under capitalism, a book which was eventually published in 1982. In this chapter, Harvey's publications are examined against the background of a growing philosophical pluralism in Anglo-American human geography.

## Philosophical Pluralism in the Geography of the Late 1970s and Early 1980s

In considering the development of Anglo-American human geography since the mid-1970s, it is necessary to note first that spatial science geography, with an integral behavioural component, has continued to exert considerable influence. Examples of this orientation include publications on model-building by Wilson (1976a, 1976b) and Thomas and Huggett (1980), an introductory text on 'scientific reasoning' written by Amedeo and Golledge (1975), and defences of the deductive scientific approach by Butterfield (1977) and Moss (1979). Systems analysis is still being utilised in geographical studies (Chapman, 1977; Bennett and Chorley, 1978; Huggett, 1980) and there has been a consolidation of research in regional science (Isard, 1975; Masser, 1976).

Liberal approaches to important contemporary social issues, largely relying upon the methods of spatial science geography, have been developed by, for example, Smith (1977) and Knox and Cottam (1981).

But criticisms of spatial science geography have also flourished (Zelinsky, 1975; King, 1976; Gregory, 1978; Smith, 1979). These have been matched by the development and consolidation of alternative approaches, giving rise to a state of philosophical pluralism in the discipline. Outstanding among alternative approaches, the Marxist perspective has been further developed in the late 1970s with *Antipode*'s future direction being identified in the editorial to a 1979 issue with rigorous Marxist analysis (O'Keefe, 1979, pp. 1-2). The Union of Socialist Geographers was formed in the United States in 1974 (Peet, 1977b, p. 256). In November of that year, the group organised a three-day seminar on Marxist geography at Clark University, which some 50 people attended.

Over the late 1970s, a fruitful dialogue developed between Anglo-American Marxist geographers and Continental Marxist sociologists, especially those associated with French urban sociology, such as Lefebvre (1970, 1972), Pickvance (1976) and Castells (1977). In March 1977, the first issue of the *International Journal of Urban and Regional Research*[1] appeared. However, its policy of publishing articles in either English or French ceased in March 1979, in order to achieve 'a satisfactory and necessary expansion of circulation' (Harloe, 1979, p. 1). All articles are now in English, although each is accompanied by an abstract in French, German and Spanish. Issues of the journal have dealt with such interdisciplinary topics as 'Urbanism and the State' (1977), Latin America (1978) and Eastern Europe (1979).

Throughout the late 1970s, Marxist geographers have expanded their critique of spatial science and liberal geography, both generally (Burgess, 1976; Anderson, 1980; Blaut, 1980) and in relation to such specific issues as location theory (Barnbrock, 1974; Walker and Storper, 1981), economic growth theory (Monstad, 1974) and urban modelling (Sayer, 1976). A Marxist perspective, utilising structural analysis and explanation, has been developed upon such topics as the housing market (Harvey, 1975c; Burgess, 1977; Le Gates and Murphy, 1981), rent theory (Walker, 1974, 1975; Harvey, 1974b; Roweis and Scott, 1978), underdevelopment and imperialism (Regan and Walsh, 1976; Lopes, 1978), unequal regional development within capitalist countries (Walker, 1978; Simon, 1980) and the plight of women (Ettorre, 1978; Whyatt, 1978) and minority cultural, ethnic and political groups (Foraie and Dear, 1978; Cooper, 1979). Some radical geographers have examined anarchist perspectives (Breitbart, 1975, 1979), reassessing in particular the work of Kropotkin (Galois, 1976; Peet, 1979) and Reclus (Dunbar, 1979). In 1977, *Radical Geography: Alternative Viewpoints on Contemporary Social Issues* was published 'to make the works of radical geographers more easily available' (Peet, 1977a, p. ix).

The phenomenological perspective has also been extended by Anglo-American human geographers, although phenomenological geographers can claim neither an organisation comparable to the Union of Socialist Geographers nor a journal of their own. Notions of the 'lifeworld' (Buttimer, 1976; Ley, 1977; Seamon, 1979a) and of 'place' (Relph, 1976a; Tuan, 1977a) are central to the phenomenological approach to geography. As Smith has noted,

> phenomenology emphasizes not the abstract conceptualizations and objective pretensions of positivism, but a more concrete concern with actual lived experience. Most obviously, it offers an alternative understanding of environment and landscape, making a clear distinction between space (the abstract object of scientific analysis) and place (the experience of objects in space, in the everyday world or *Lebenswelt*) (Smith, 1979, p. 365).

Phenomenological criticism of positivist methods has continued (Seamon, 1975; Relph, 1976b; Ley, 1978, 1981) while research has proceeded, applying the methods of participant observation and experiential fieldwork (Tuan, 1975; Rowles, 1978; Seamon, 1979a). One important line of investigation in social geography has been based upon the phenomenological sociology of Alfred Schutz (Duncan and Duncan, 1976; Ley, 1977, 1978).

Considerations of existentialism first appeared in the geographical literature in the early 1970s (Samuels, 1971; Tuan, 1972; Lee, 1974) but little has been done from the existentialist perspective since then (but see Van Paassen, 1976, and Samuels, 1978, 1981). In 1978, a number of essays written from the phenomenological and existential viewpoints were included in a publication entitled *Humanistic Geography: Prospects and Problems* (Ley and Samuels, 1978). Entrikin (1976) has argued that humanistic geography is best understood as a form of criticism of positivist geography, arising out of a critical humanist tradition, whereas Tuan (1976, 1977b) has viewed humanistic geography as a subfield of geography, dealing with the overlap between geography and the humanities. Hence the study of the relationships between art and geography is a humanistic study (Tuan, 1978; Relph, 1979; Pocock, 1981).

A variety of idealist approaches have also been advocated by a small but significant number of geographers over this period. In 1974, Guelke first suggested that an idealist alternative in human geography be developed, based upon Collingwood's (1956) work. Guelke has since applied idealism to the study of frontier settlement in early Dutch South Africa (Guelke, 1976) and written further articles supporting an idealist geography (Guelke, 1977a, 1977b, 1978, 1981). Neo-Kantianism, which has an idealist character, has been discussed by Berdoulay (1976) in relation to French geography, and by Entrikin (1977) in relation to Ernst Cassirer's philosophy of space. Livingstone and Harrison (1981) have also advocated a return to the spirit of neo-Kantianism, whereas Olsson (1975, 1980) and Marchand (1974, 1978, 1979) have explored a type of dialectical idealism.

That pluralism in geography was becoming accepted within the discipline at the end of the 1970s was confirmed by the publication in 1979 of *Philosophy in Geography*, a collection of articles written from diverse points of view (Gale and Olsson, 1979a). In 1981, a text appeared for advanced undergraduate and graduate courses in contemporary philosophical themes in geography (Harvey and Holly, 1981a), illustrating the growing recognition of pluralism at this level of teaching. But Bird has also identified a deeply rooted polarity within geography, reflected in such contrasts as those between *erklaren* ('explanation in the natural sciences') and *verstehen*, ('understanding in the human sciences'), between a categorical and a dialectical paradigm and between 'empirical and analytic' and 'historical and hermeneutic' forms of knowledge (Bird, 1979, p. 119). Within human geography, this polarity has given rise not only to the contrast between the positivist and (structuralist) Marxist approaches on the one hand and the humanist and phenomenological perspectives on the other but also to tensions within each approach or perspective.

Harvey's work over this period was within the Marxist philosophical tradition, utilising a structuralist method of analysis and explanation. He had rejected positivism, phenomenology and idealism in 1972 (Harvey, 1972c, pp. 10-11). However, he encouraged political, ideological and scientific pluralism in geography, mainly, it appears, to provide the opportunity for Marxist geographers to continue their work without fear of repression (Harvey, 1977c, p. 407).

## David Harvey's Geography, 1974-1981

Between 1974 and 1981, Harvey published ten articles on urbanisation under advanced capitalism, based on his research in Baltimore. In two of these articles, he dealt specifically with aspects of Marx's view of society, namely, his theory of the accumulation of capital and of the capitalist State. In this programme of research, Harvey was concentrating upon the demystification of the capitalist urban process, demonstrating that the city was being manipulated as a tool for solving the crises of a capitalist economy. In two further papers published towards the end of the decade, Harvey used his research to show how urban and regional planners contributed to the maintenance of the capitalist economic and social order and to open up dialogue with urban sociologists in the Chicago School tradition, and in a third article he recounted the class struggle associated with the building of a Parisian church in the nineteenth century. Since 1978, Harvey has also written a number of short articles for the *Baltimore Sun* and *The Progressive* on issues such as academic freedom and revolutionary struggle in Central America. In this chapter the opportunity will also be taken to examine Harvey's exchanges with Berry and Carter over academic freedom and philosophical and political pluralism.

### *The Housing Market, Financial Institutions and Social Classes*

In April, 1974, *FHA Policies and the Baltimore City Housing Market*, written by Chatterjee, Harvey and Klugman, was published by the Baltimore Urban Observatory. In the report, the authors examined the role of the Federal Housing Administration (FHA) and other financial institutions in the Baltimore housing market. They identified a conflict between the FHA's social welfare and investment roles (Chatterjee, Harvey and Klugman, 1974, p. 15.1) but they also pointed out that many of the problems associated with its functioning arose from 'deeper problems of a systematic nature', which could not be solved by internal adjustments alone (Chatterjee *et al.*, 1974, p. 6.2). However, they drew back from explicitly criticising the capitalist context of the housing market.

In two articles published in 1974, Harvey combined his previous interest in the rent concept with a more explicitly Marxist analysis of the Baltimore housing market, concentrating upon the roles of financial institutions and social classes. The first article, written with Chatterjee, was entitled 'Absolute Rent and the Structuring of Space by Governmental and Financial Institutions'. It was published in *Antipode* and began by posing two questions: 'How do the macro and micro features of housing markets relate to each other in theory and in practice?' and 'How is absolute rent realized in the housing markets of large metropolitan areas?' (Harvey and Chatterjee, 1974, p. 22). In examining the first question, the authors briefly outlined Ollman's (1971 and 1972) view of the methodology of internal relations, pointing out that it defined as the immediate object of inquiry 'the *processes* of structuring and transformation together with the determinate structures that mediate these processes' (Harvey and Chatterjee, 1974, p. 22). Harvey and Chatterjee went on to analyse the Baltimore housing market utilising this 'methodological stance' (Harvey and Chatterjee, 1974, p. 22).

National housing policies in the United States were designed to maintain the existing capitalist structure of society by 'facilitating economic growth and capitalist accumulation, eliminating cyclical influences and defusing social discontent' (Harvey and Chatterjee, 1974, p. 23). Housing was one of the tools used by government to implement these policies which, according to Harvey and Chatterjee, were transmitted to the local level by the actions of financial institutions. As Figure 4.1 illustrates, the immediate determinant of housing choice was not income, as microeconomic models assumed, but rather the ability to obtain credit and mortgages from financial and governmental institutions (Harvey and Chatterjee, 1974, pp. 23-4).

Harvey and Chatterjee demonstrated that the eight housing submarkets in Baltimore were distinguished from one another by variations in the activity of financial institutions, although this was only possible in some instances by overgeneralising the data which they used. For example, they concluded that the 'ethnic areas' were dominated by small community and neighbourhood State Savings and Loans institutions (S and Ls) whereas the data showed that, while State S and Ls accounted for 43.2 per cent of the financing, cash and private transactions accounted for 45.4 per cent (Harvey and Chatterjee, 1974, pp. 25, 27).

Harvey and Chatterjee suggested that it was within the 'geographical structure' of the submarkets that individual households made housing choices which tended to conform to and reinforce the structure. As illustrated in Figure 4.1, transformation of the geographical structure could take place through the operations of speculators, the changing profitability of 'landlordism' and changing governmental and institutional policies (Harvey and Chatterjee, 1974, pp. 25-32).

On the subject of the realisation of absolute rent, the theoretical aspect of which Harvey had examined in Chapter 5 of *Social Justice*

Figure 4.1: The Structuring and Transforming Forces Shaping Urban Housing Submarkets (based upon Harvey and Chatterjee, 1974, 22-32 and Harvey, 1974b, 243-50).

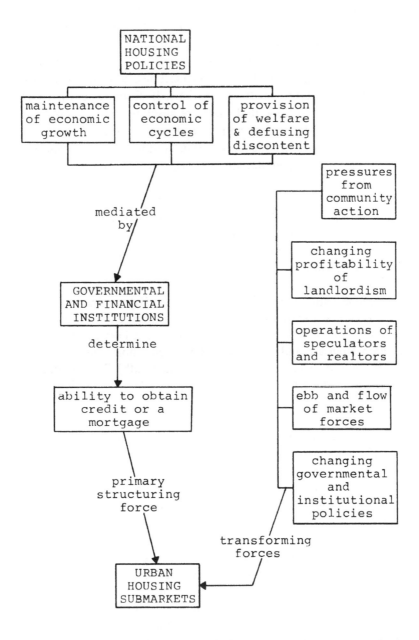

*and the City*, Harvey and Chatterjee pointed out that this form of rent implied 'class monopoly power of some sort'.

> By a "class monopoly" we mean a class of producers (or consumers) who have power over a class of consumers (or producers) in a situation of structured scarcity. We have first to define the basis for such class monopoly power in the housing market . . . Through the structuring activity of governmental and financial institutions, urban space is differentiated into specific submarkets. If absolute rent is to be realized we have to show that there are absolute limits of some sort operating over different segments of the housing market. These absolute limits can be set by the joint attributes of housing, of financiers, of housing suppliers and of consumers (Harvey and Chatterjee, 1974, pp. 32-3).

Thus, the geographical and social structures of the housing submarkets were seen to be the preconditions for the realisation of absolute rent (see Figure 4.2). Harvey and Chatterjee gave the example of the inner-city submarket, where

> a low-income and uncreditworthy population [is] confined to tenant occupancy in a situation where landlords . . . must extract a high rate of return out of bad housing. Absolute rent is here gained by the landlord from a population that has no other choice (unless, in the Baltimore case, it migrates back to the rural South). This population is trapped in the structure of absolute space (Harvey and Chatterjee, 1974, p. 34).

The authors concluded that they had shown that a great deal may be learned if geographers adopted 'a methodology appropriate for understanding society as a totality fashioned through a structured set of internal relations' (Harvey and Chatterjee, 1974, pp. 34-5).

Harvey continued his consideration of the concept of rent in relation to class monopoly power in an article entitled 'Class-monopoly Rent, Finance Capital and the Urban Revolution' (Harvey, 1974b). No mention was made of Marx's method in this article. Harvey began by noting that value arose out of production and was realised in consumption. An elaborate social structure, a structure of social institutions to co-ordinate individual and group activities and a physical infrastructure were all needed for production and distribution to take place. 'Transfer payments' had to be made out of the value produced in order to maintain

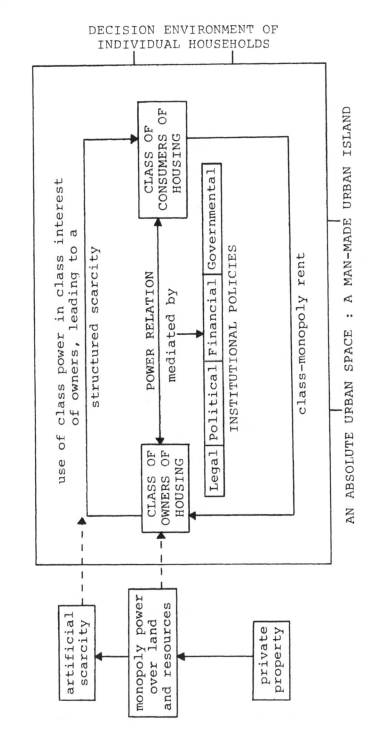

Figure 4.2: The Structure of an Urban Housing Submarket (based upon Harvey and Chatterjee, 1974, 23-5 and 32-4 and Harvey, 1974b, 240-9).

them. As illustrated in Figure 4.1, Harvey viewed rent as a transfer payment realised through the monopoly power over land and resources conferred by the institution of private property (Harvey, 1974b, p. 240). The type of rent that Harvey was considering did not fall easily into either the absolute or monopoly rent categories that he had examined in Chapter 5 of *Social Justice and the City*. He therefore coined the term 'class-monopoly rent', defining class, as in the previous article (Harvey and Chatterjee, 1974, p. 36, note 18), as 'any group that has a clearly defined common interest in the struggle to command scarce resources in society' (Harvey, 1974b, p. 241, note). Harvey pointed out that Marx (1967, Volume 3, pp. 194-5) had himself used the notion of class-monopoly power.

> Class-monopoly rents arise because there exists a class of owners of "resource units" — the land and the relatively permanent improvements incorporated in it — who are willing to release the units under their command only if they receive a positive return above some arbitrary level (Marx, 1967, Chapter 45). As a class these owners have the power always to achieve some minimum rate of return (Harvey, 1974b, p. 241).

Harvey suggested that a power struggle would ensue between low-income tenants and their landlords. Whereas the landlords would be seeking maximum rents, the tenants would attempt to use their political power to impose minimum housing standards or rent controls. The landlords might then attempt to block these moves in order to maintain their income. The rate of return set through the working out of this conflict was best interpreted as a class-monopoly rent (Harvey, 1974b, p. 242). Elsewhere in the housing market, suburban middle-income and upper-income groups were confronted by a class of speculator-developers who might attempt to manipulate zoning decisions to gain greater returns.

> If the speculator-developer can persuade upper-income groups of the virtues of a certain kind of housing in a particular neighbourhood, gain complete control over the political process, and so on, then the advantage lies with the speculator-developer. If consumers are unimpressed . . . and have firm control over the political mechanisms for land-use regulation, . . . then the class-monopoly power of the speculator-developers will be contained (Harvey, 1974b, p. 243).

Harvey asserted that class-monopoly rent appeared inevitable in capitalist housing markets where a hierarchical structure existed, rent being passed from a low-income tenant to a landlord to a speculator-developer to a financial institution at the top (Harvey, 1974b, p. 243). This structure was necessary in order that the government could co-ordinate the housing market through financial institutions to avoid economic crises (Harvey, 1974b, p. 244). Harvey then repeated in more detail the analysis in Harvey and Chatterjee (1974, pp. 24-30) of the structure of the Baltimore housing submarkets with respect to the type of institutional involvement (Harvey, 1974b, pp. 245-9).

Harvey characterised the city as consisting of man-made islands of absolute space on which class-monopolies produced absolute scarcities (Figure 4.2). These absolute spaces were constructed primarily by the activity of financial institutions in the housing market (Harvey, 1974b, p. 249). Residential differentiation was thus not the result of social eco-logical processes, consumer preferences, utility-maximising behaviour and so on, although such features helped to maintain the island-like structure. 'There is a deeper process at work', asserted Harvey. 'Financial institutions and government manage the urbanization process to achieve economic growth, economic stability and to defuse social discontent' (Harvey, 1974b, p. 250). Harvey believed that the processes he had isolated could be generalised to all advanced capitalist nations although the particular manifestations of these processes could not (Harvey, 1974b, p. 250).

Turning his attention to the role of 'class interest' in the realisation of class-monopoly rent, Harvey distinguished between subjective classes, which described the consciousness which groups had of their position within a social structure, and objective classes, which described the basic division within capitalism between producers and appropriators of surplus value (Harvey, 1974b, p. 250). Community-based conflict tended to arise from confrontation between subjective classes, not objective classes. 'Marx thought that large concentrations of popula-tion would heighten class-awareness', Harvey commented. 'But under urbanization class-consciousness appears to have become fragmented' (Harvey, 1974b, p. 251). Community conflict tended to be parochial, community being pitted against community, so that the average condi-tion of communities was not altered. Both non-parochialist community conflict and work-based conflict led to confrontation with the power of 'finance capital'. In the housing market, for example, financial institu-tions were to be found at the top of the financial hierarchy, whereas in the industrial enterprise, financial institutions exercised powerful ex-ternal control (Harvey, 1974b, pp. 251-2).

Finance capital created a range of subjective classes in its creation of
new social wants and needs, Harvey believed. The creation of housing
consumption classes in middle and upper-income groups, for instance,
was achieved through the activities of developers financed by various
institutions. Such subjective classes, occupying certain housing submarkets, functioned to reproduce the social relations of labour under
capitalism (Harvey, 1974b, pp. 252-4).

As a result of his findings, Harvey suggested that Marx's theory of
surplus value ought to be embedded in a general theory of exploitation
in order to take account of the wide-ranging control of finance capital
over the totality of production, circulation and the realisation of value
in society (Harvey, 1974b, p. 253). Harvey further suggested that objective classes be defined in terms of the totality of the production process which included the immediate production of value, the production
of new modes of consumption and of new social wants and needs, the
production and reproduction of labour power and of the social relations
of capital (Harvey, 1974b, p. 254).

Harvey concluded that urbanism had been transformed from an expression of the production needs of the industrialist to

> an expression of the controlled power of finance capital, backed by
> the power of the State, over the totality of the production process
> . . . It seems, however, that the finance form of capitalism, which
> has emerged as a response to the inherent contradictions in the com-
> petitive industrial form, is itself unstable and beset by contradictory
> tendencies . . . The perpetual tendency to try to realize value without
> producing it is, in fact, the central contradiction of the finance form
> of capitalism (Harvey, 1974b, p. 254).

Whereas literature on the Third World had dealt explicitly with the
relationships between urbanisation, economic growth, capital accumulation and the structuring of social classes, Harvey noted that such
relationships were ignored in the context of the advanced capitalist
countries. 'To make these relationships more explicit is an urgent task
to which this paper seeks to make a modest beginning' (Harvey, 1974b,
p. 254).

In an article published in 1975, Harvey continued to explore the
relationships between urbanisation, economic growth, the accumulation
of capital and the structuring of social classes. He used his previous
analysis of the Baltimore housing market to exemplify his theoretical
points. Entitled 'Class Structure in a Capitalist Society and the Theory

of Residential Differentiation', the article was published in a collection of essays marking the fiftieth anniversary of the establishment of the Department of Geography at the University of Bristol (Harvey, 1975a). In it, Harvey outlined the relationships between Marxism as a general social theory and the theory of residential differentiation, utilising the 'Marxian method . . . founded in the philosophy of internal relations' (Harvey, 1975a, p. 355).

Relying upon Marx, Giddens (1973) and Poulantzas (1973), Harvey identified three kinds of forces of class structuration, that is, three forces shaping social differentiation within a population. The primary force was that arising out of the 'power relation between capital and labour'[2] (Harvey, 1975a, p. 362). Harvey noted that the two-class model presented by Marx in *Capital*, Volume 1, was but an 'analytic construct' used to lay bare the exploitative character of capitalist production (Harvey, 1975a, p. 357). It was not meant as a description of an actual class structure, although Marx often used the dichotomous model as if it had empirical content and insisted that socialism would emerge after a class struggle between capitalists and the proletariat.

> The analytic constructs of *Capital* consequently become normative (ought to be) constructs in his programmatic writings. And if actual class struggle crystallised around the capital-labour relation, then both the analytic and normative constructs would come to take on an empirical validity as descriptions of actual social configurations (Harvey, 1975a, p. 357).

Harvey identified a variety of secondary forces of class structuration, derivative of the primary force, which encouraged social differentiation along lines defined by the division of labour, consumption patterns and lifestyle, and the manipulated projections of ideological and political consciousness (Harvey, 1975a, p. 362). 'Residual forces', reflecting social relations established in preceding or geographically separate but subordinate modes of production, were viewed as the third kind of force shaping the structure of social classes (Harvey, 1975a, p. 362).

Harvey fashioned four hypotheses connecting residential differentiation with the social structure he had just examined. In the first, he stated that residential differentiation had to be interpreted in terms of the reproduction of the social relations of capitalism. Secondly, he hypothesised that residential areas were distinctive milieus from which people derived their values, expectations, consumption habits, states of consciousness and so on. Harvey's third hypothesis was that the fragmentation of society into distinctive communities fragmented class consciousness, thereby frustrating the transition, through class struggle, from capitalism to socialism. Fourthly, Harvey believed that patterns of residential differentiation reflected and incorporated many of the contradictions of capitalism (Harvey, 1975a, pp. 362-3).

In the subsequent general discussion in support of these hypotheses, Harvey argued that suburbanisation was the creation of the capitalist mode of production in an attempt to maintain effective demand, thereby facilitating the accumulation of capital. The capitalist mode of production created 'a distinctive group of white collar workers . . . imbued with the ideology of competitive and possessive individualism' to the extent that suburbanisation had become their way of life (Harvey, 1975a, p. 367).

> We can, thus, interpret the "preference" for suburban living as a created myth . . . But like all such myths, once established it takes on a certain autonomy out of which strong contradictions may emerge. The American suburb, formed as an economic and social response to problems internal to capitalist accumulation, now forms an entrenched barrier to social and economic change. The political power of the suburbs is used conservatively . . . A phenomenon created to sustain the capitalist order can in the long run work to exacerbate its internal tensions (Harvey, 1975a, p. 367).

Thus, Harvey continued, to maintain its own dynamic, capitalism is forced to disrupt and destroy what it initially created to preserve itself (Harvey, 1975a, pp. 367-8). Harvey therefore concluded that

> instead . . . of regarding residential differentiation as the passive product of a preference system based in social relationships, we have to see it as an integral mediating influence in the processes whereby class relationships and social differentiations are produced and sustained (Harvey, 1975a, p. 368).

Harvey further commented upon the theory of residential differentiation in a review of Foster's (1974) *Class Struggle and the Industrial Revolution – Early Industrial Capitalism in Three English Towns* (Harvey, 1975f). He identified an implicit thesis growing out of Foster's argument regarding the role of residential differentiation in relation to occupational structure and mass-based class consciousness.

I infer from chapter 5, for example, that the defusing of mass-based class consciousness entailed, and perhaps even involved, a growing occupational residential segregation. Here, perhaps, lie the roots of a contemporary phenomenon — residential differentiation which reflects the parochialisms of occupation rather than the politics of class struggle (Harvey, 1975f, p. 110).

Harvey regarded Foster's book as a 'pioneering classic' mainly because it successfully fused a discussion of broad issues of social and economic change, such as the dynamics of capitalist society, with 'the details of what happened on the ground' (Harvey, 1975f, p. 109).

## The Political Economy of Urbanisation

In 1975, an article by Harvey entitled 'The Political Economy of Urbanization in Advanced Capitalist Countries: the Case of the United States' was published in *The Social Economy of Cities*, a collection of papers edited by Gappert and Rose (Harvey, 1975c). Harvey's contribution had been circulated in mimeographed form at the eighth World Congress of Sociology in Toronto in 1974 (Mingione, 1977, p. 106, note 4). Harvey acknowledged that it was part of a general investigation of the political economy of urbanisation in advanced capitalist countries which had begun with *Social Justice and the City* and had continued in a series of articles (Harvey and Chatterjee, 1974; Harvey, 1974b, 1975a). In the first part of the article, Harvey established certain broad relationships between urbanisation and economic growth processes. He pointed out for example, that between a fifth and a quarter of gross domestic product went into fixed capital formation in advanced capitalist countries. Between a half and two-thirds of this typically went into building construction, the rest into transport and machinery (Harvey, 1975c, pp. 120-1). Harvey then noted how investment in construction had decreased in importance, especially relative to machinery. This, according to Harvey, indicated that

> much of the urban infrastructure in the U.S. was created prior to 1920 and that the fixed assets created by then are still heavily drawn upon ... In general, it seems that there is a long term shift from urban infrastructure ... to producers' durables in the production sphere and a parallel shift from urban infrastructure ... to consumers' durables ... in the consumption sphere (Harvey, 1975c, pp. 122-3).

Despite this, investment in the built environment had a significance far beyond the direct investment absorbed, for important multiplier effects were generated, urban infrastructure had a long life span and efficiency in the urban infrastructure decreased production costs (Harvey, 1975c, p. 124).

Harvey went on to examine the role of financial institutions in relation to urbanisation. He viewed these institutions as a superstructural form which had evolved in response to the need to maintain the rate of capital accumulation in a capitalist mode of production. If this was so, then

> it can be anticipated that the urban problems with which we are all too familiar (inner-city decay in American cities, sky-rocketing land and housing prices in European and Canadian inner-cities, speculative developer activity, suburban sprawl, pollution and congestion, the lack of balance between housing and employment opportunities, and the like) are all surface manifestations of deep structural problems in such [advanced capitalist] societies (Harvey, 1975c, p. 127).

Harvey's studies of the housing market led him to believe that the relationship between the economic basis, the financial superstructure and urbanisation was in need of systematic investigation. This suggested that there was a need to examine the financial superstructure of the United States in relation to housing in the light of what he called 'the Marxian posture', that is, the view that urbanisation was a superstructural form (Harvey, 1975c, p. 128).

The overwhelming impression created by a study of financial structure in the United States was, according to Harvey, one of 'a chaos of private activity under an incoherent and arbitrarily constructed umbrella of government regulation' (Harvey, 1975c, p. 129). Despite this, a coherent investment process took place, through which funds were channelled into certain aspects of the housing sector (Harvey, 1975c, pp. 130-1). Harvey identified single-family homeownership as being particularly favoured by the policies and practices of private, federal and local governmental financial institutions in the United States, as represented in Figure 4.3. In the 1930s, for instance, a number of federal institutions were created to ensure the stability of the financial superstructure while channelling funds to homeowners. In 1968, one of these institutions, the FHA, was transformed so that it could become 'a vehicle for promoting homeownership among the poor and near-poor' (Harvey, 1975c, p. 133). Tax deductions on mortgage interest

Figure 4.3: Federal Policies and the Urbanisation Process in the United States, 1930s to 1970s (based upon Harvey, 1975c, 129-36).

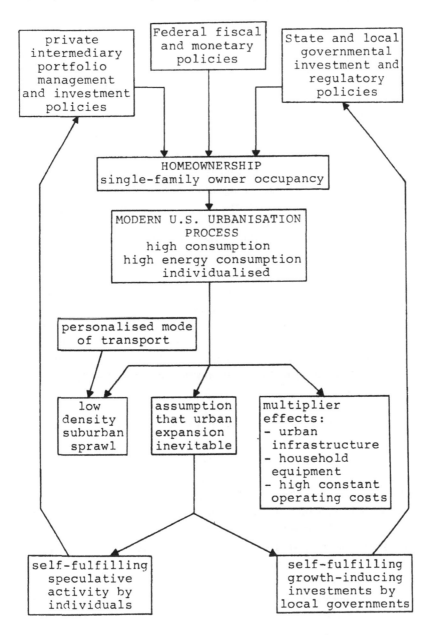

and property tax payments cost the federal government $US six billion in 1972. Because of these and other factors, housing construction experienced a 'remarkable' boom after World War II, which gave rise to the modern urbanisation process in the United States. This urbanisation was characterised by low-density suburban sprawl, strong multiplier effects and activity on the part of both individuals and local governments, which reinforced the widely held assumption that urban expansion was inevitable. These relationships are set out in Figure 4.3, which also illustrates Harvey's view that relatively low land costs, increasing geographical mobility and social and economic conflict in the inner cities have all compounded these trends (Harvey, 1975c, p. 135).

Harvey went on to point out that rising land prices, no-growth suburban movements, increasing congestion and pollution costs, fiscal inequality among local governments and the rising costs of energy were beginning to force a serious reconsideration of the benefits of the urbanisation process in the United States. In addition, the general levels of indebtedness attached to urban infrastructure had risen dramatically.

> Residential debt as a proportion of total private and public national debt rose from 9.5% in 1947 to 23.7% in 1972, while expressed as a ratio to GNP it has now risen to levels comparable to those experienced in the crisis conditions of the 1930s . . . In general, residential debt has become the largest single component underpinning the viability and security of the financial superstructure (Harvey, 1975c, p. 136).

Harvey believed that the stability of both the social structure and the financial superstructure were largely dependent upon the health of the residential mortgage market. In other words, the stability of the financial system, the ability of surplus capital to find outlets in real estate development, and the ability of individuals to bear the burden of long-term indebtedness were but separate facets of the same process (Harvey, 1975c, p. 137). Furthermore, the structure of debt-encumbrance was such that it was predicated on future economic growth so that 'future urban growth may be essential to secure indebtedness incurred in creating past urban growth' (Harvey, 1975c, p. 137). Thus, Harvey concluded, the American city was no longer a workshop for the production of goods and services but was rather designed to stimulate consumption in order to prevent the return of severe depression conditions (Harvey, 1975c, p. 139).

It was at this point that Harvey turned his attention to the Baltimore housing market, illustrating how the creation and transformation of submarkets were due to the need to meet the demands of capital accumulation. He examined three case studies, namely, West Baltimore and the use of the land-instalment contract, Northeast Baltimore and the activity of the FHA, and the ethnic submarkets and the evolution of national financial policy. Harvey's aim was to show how the structure of financing had shaped residential differentiation. West Baltimore, for example, was originally a white middle-income area. In the 1960s, blacks, who had been denied funds from 'white establishment' financial institutions, were financed into West Baltimore by landlord-speculators who, relying upon funds from mortgage bankers, were able to buy and sell housing at some profit. A speculator would typically pick up a house from a fleeing white family for around $US 7,000 and then sell it to a black family for $US 12,000 or so, as well as providing mortgage finance (at an interest rate greater than that for which it was initially obtained). The speculator would retain title to the house until the mortgage was paid up.

> A community group ... documented over 4,000 land-instalment contract transactions in Baltimore City in the early 1960s and the majority of these were in West Baltimore. One organization bought 1,768 houses for $10.8 million and sold 742 of them on the land-instalment contract for $9.4 million (the rest going into the rental inventory). The average purchase price for the speculator on the houses finally sold was $6,868 compared to final average sales price of $12,706. The net mark-up, called the "black tax", became the center of controversy in Baltimore during the 1960s, as did the very high rate of induced turnover in the housing stock (Harvey, 1975c, p. 145).

Harvey believed that the urban riots of the 1960s were partly a response to this form of exploitation. Public pressure on the financial institutions not to finance speculative activity in the housing market and the 'public pillorying' of speculators who used the land-instalment contract eventually led to the end of this speculative activity in West Baltimore. However, by 1969, West Baltimore was a black residential area, many of the previous inhabitants having moved to the suburban fringe where land speculators, developers and construction interests were actively investing with resources from the financial institutions. Communities were thus disrupted and exploited by the financial superstructure as it sought to provide a dynamism to the urbanisation process which matched the dynamism of capitalist accumulation in general (Harvey, 1975c, pp. 144-7).

Long-submerged contradictions within the urbanisation process were beginning to surface in the mid-1960s and Harvey emphasised one which involved the delicate balance between writing off the value of capital assets in the city in order to stimulate effective demand and yet preserving the value of debts incurred by financial institutions (Harvey, 1975c, p. 157).

> And herein lies the contradiction. If the flight to the suburbs continues at its present pace ... then there are bound to be serious repercussions within the national financial structure. If the flight does not continue there may be equally serious repercussions for the maintenance of effective demand and a return to the underconsumption problems of the 1930s (Harvey, 1975c, p. 158).

Harvey finally concluded by outlining 'the substantive thesis' that he had sought to document in the article.

> There is an intimate connection between financial superstructure and the shape and form taken by the urbanization process. Since the financial superstructure has largely been fashioned as a response to problems in the sustained accumulation of capital and in particular to crises in that process, the financial superstructure mediates the relationship between the main dynamic of sustained capital accumulation, on the one hand, and the urbanization process, on the other. Within these relationships we can see a variety of conflicts, tensions, and contradictions. The evidence is incontrovertible that urbanization manifests and perhaps contributes to many of the contradictions implicit in a dynamic capitalist mode of production (Harvey, 1975c, pp. 160-1).

This article was deemed to be significant enough to be translated into French and published in *Espaces et Sociétés* in 1976. Harvey was thus becoming widely known as an authority on the political economy of urbanisation.

In November, 1974, Harvey presented a paper at the seminar on Marxist geography at Clark University organised by the then fledgling Union of Socialist Geographers. Entitled 'Some Remarks on the Political Economy of Urbanism' (Harvey, 1975b), it was based upon the article just examined (Harvey, 1975c). Perhaps as a result of the more informal context in which it was presented, the paper, as reported in *Antipode*, revealed more than the published article with regard to Harvey's intentions.

> The most important revelation I gained from reading Marx is that there is much mystification around; most of what academics do is designed to hide from us what is really happening. The structure of knowledge leads us to think about capitalism in pro-capitalist terms and not about alternatives to it. Therefore we have to begin a process of demystification and build a critical theory which will reveal to us what is really happening.
>
> This suggests an initial imperative: in order to change the world we have to understand it. Such understanding is an intellectual activity. There is a simultaneous imperative: to design and discover ways of acting which begin the process of transformation (Harvey, 1975b, p. 55).

In Harvey's view, the educational activities of the Union of Socialist Geographers must relate local issues to 'the total economy' in order to develop class consciousness. Socialist geographers had to identify, analyse and offer alternatives to capitalism (Harvey, 1975b, p. 58).

In 1974, Harvey presented a paper at a symposium celebrating a half century of geography at Michigan University. Entitled 'Government Policies, Financial Institutions and Neighbourhood Change in U.S. Cities', it drew heavily upon 'The Political Economy of Urbanization in Advanced Capitalist Countries: the Case of the United States' (Harvey, 1975c) and was published in 1977 in a collection of the proceedings of the Michigan symposium (Harvey, 1977a) as well as in Harloe's *Captive Cities: Studies in the Political Economy of Cities and Regions* (Harvey, 1977b). In it, Harvey emphasised the disruptive effects of changing national economic policies upon housing markets. He concluded that

> The contemporary history of residential differentiation in any city in the United States (and Baltimore has been used as an example here) shows that communities are disrupted, populations moved (often against their will) and the whole structure of the city altered as the urbanization process, co-ordinated in its major outlines through the mediations of governmental and financial structures, is utilized as a vehicle to sustain an effective demand for product. An accelerating rate of "planned obsolescence" in our cities appears as a necessary

evil to feed the dynamics of capital accumulation and growth within the U.S. economy (Harvey, 1977a, pp. 312-3).

## *Accumulation, Class Struggle and the State*

The third article of Harvey's published in 1975 was entitled 'The Geography of Capitalist Accumulation: a Reconstruction of the Marxian Theory' (Harvey, 1975d).[3] It was published in *Antipode* and in it Harvey examined Marx's theory of the accumulation of capital, a topic important to Harvey's work because he had recognised capital accumulation as the fundamental process shaping capitalist society (Harvey, 1975c, p. 127). In a footnote, Harvey mentioned that the article was partly taken from 'Chapter 5 of a forthcoming book on urbanization under capitalism' (Harvey, 1975d, p. 9, note), the first indication that Harvey was planning to publish a third book.

For Marx, capitalist production was characterised by 'accumulation for accumulation's sake, production for production's sake' (Harvey, 1975d, p. 9).

> Accumulation is the engine which powers growth under the capitalist mode of production. The capitalist system is therefore highly dynamic and inevitably expansionary; it forms a permanently revolutionary force which continuously and constantly reshapes the world we live in (Harvey, 1975d, p. 9).

The process of the accumulation of capital did not stem from the inherent greed of the capitalist, according to Harvey. Rather, it was an 'imperative'. He quoted Marx (1967, Volume 1, p. 592) as writing that

> the development of capitalist production makes it constantly necessary to keep increasing the amount of capital laid out in a given industrial undertaking, and competition makes the immanent laws of capitalist production to be felt by each individual capitalist, as external coercive laws. It compels him to keep constantly extending his capital in order to preserve it, but extend it he cannot, except by means of progressive accumulation (Harvey, 1975d, p. 9).

Harvey indicated that the progress of the accumulation of capital depended upon three factors, namely, the existence of a surplus of labour power, access to further supplies of the means of production (machines, raw materials, physical infrastructure, and so on), and the existence of a market to absorb the increasing quantities of goods produced (Harvey,

1975d, p. 9). But the form of any of these three could also come to constitute a barrier to further accumulation. Harvey gave the example of a realisation crisis, when commodities were on the market but there were no purchasers. Commodities tended to be produced to the limit of the productive forces without regard to the limits of the market. Wages were kept down to maximise profit but this restricted the purchasing power of consumers. The result, a realisation crisis, was relative over-production in terms of effective demand (defined as 'need backed by ability to pay') (Harvey, 1975d, p. 10). Harvey believed that a realisation crisis could also appear as a crisis of underconsumption or of the overproduction of capital. The resolution of a realisation crisis lay in expanding effective demand by the penetration of capital into new spheres of activity, the creation of new social wants and needs, encouraging population growth or geographically extending the market (Harvey, 1975d, p. 11).

Periodic crises were endemic to capitalism. The tendency to over-accumulate was therefore inherent in capitalist production, powered by the imperative to accumulate. By continually seeking to extend its activity to resolve such crises, capitalism simply increased their geographical extension (Harvey, 1975d, p. 11). If geographical concentration and expansion were to occur alongside of the continued accumulation of capital, there would have to be a decreasing cost in the movement of commodities ('the overcoming of spatial barriers') and a reduction in the time involved in the movement of commodities ('the annihilation of space by time') (Harvey, 1975d, p. 12). This, according to Harvey, was the basis of Marx's location theory. A distinct spatial structure evolved, that is, a certain pattern and density of production and population, which, Harvey pointed out, eventually could become a barrier to further accumulation. Space for an increasing level of accumulation, that is, for a more efficient concentration of production and population, required the destruction of past capital investments. Harvey viewed this as a contradiction which led to periodic crises in fixed capital investment and to periodic reshapings of the geographic environment (Harvey, 1975d, pp. 12-13). Thus, 'the landscape which capitalism creates is . . . seen as the locus of contradiction and tension', the crises of capitalism leading to 'the dialectical transformation of geographical space' (Harvey, 1975d, p. 13).

In this the Marxian approach is fundamentally different to that typical of bourgeois economic analysis of locational phenomena. The latter typically specifies an optimal configuration under a specific set

of conditions and presents a partial static equilibrium analysis . . .
Dynamics never get much beyond comparative statics (Harvey,
1975d, p. 13).

Examining Marx's writings on foreign trade, Harvey pointed out that
Marx's theory indicated that capitalism could escape its own contradictions only through expanding. Expansion was simultaneously '*in-
tensification* (of social wants and needs, of population totals, and the
like) and *geographical extension*' (Harvey, 1975d, p. 17), although
Marx's theory did not predict where, when and how these would occur
in any specific situation. Thus any failures in Marx's historical predictions were, according to Harvey, failures of historical analysis, not
failures of theory (Harvey, 1975d, p. 16).

For Harvey, a theory of imperialism was a theory of history and he
believed that Marx's concern was not with such an historically specific
analysis. Harvey evaluated such Marxist theories of imperialism as those
of Lenin (1965), Luxemburg (1951), Baran (1957), Frank (1969) and
Amin (1973), pointing out how they drew upon only one or two facets
of Marx's theory of accumulation (Harvey, 1975d, pp. 18-19).

Harvey concluded the article by advocating that Marx's theory of
capital accumulation on an expanding and intensifying geographical
scale be 'forced' into an intersection with 'materialist investigations of
actual historical configurations' (Harvey, 1975d, p. 20).

We have to learn, in short, to complete the project which Marx
underscores at the beginning of Volume 3 of *Capital* — we have to
bring a synthetic understanding of the processes of production and
circulation under capitalism to bear on capitalist theory and 'thus
approach step by step the form which they assume on the surface
of society' (Harvey, 1975d, p. 21).

The significance of 'The Geography of Capitalist Accumulation: a
Reconstruction of the Marxian Theory' lay in the exploration conducted
by Harvey into the process he believed to structure the form of urbanisation under capitalism. Although he examined Marx's theory of the
accumulation of capital in a broader context than his research on urbanisation required, Harvey later indicated that accumulation was one of
the two themes upon which his understanding of the capitalist urbanisation process was based.

In 1976, Harvey published two articles extending certain theoretical
aspects of his previous work on urbanisation under advanced capitalism.

In the first, he dealt with class struggle around the built environment (Harvey, 1976a) whereas in the second he examined the Marxian theory of the State (Harvey, 1976b). Both articles were acknowledged as drawn from a forthcoming book on urbanisation in advanced capitalist countries. In a third paper, published in 1978, Harvey encouraged urban and regional planners to critically review their role as 'instruments' of the capitalist State in its attempt to reproduce the capitalist social order and maintain the accumulation of capital (Harvey, 1978a).

Entitled 'Labor, Capital, and Class Struggle around the Built Environment in Advanced Capitalist Societies', the first article appeared in a journal called *Politics and Society*, indicative of the interdisciplinary character of Harvey's research and his continued interaction with scholars outside of the confines of geography. In the article, Harvey sought to establish a theoretical framework for understanding a facet of class struggle under advanced capitalism, that which took place around the built environment. He defined the built environment as a totality of physical structures which included houses, roads, factories, educational facilities and parks. Harvey then identified four groups with interests in the built environment, namely, a 'faction of capital' seeking rent, a 'faction of capital' seeking interest and profit, 'capital "in general"' seeking an outlet for surplus capital and a means for the production and accumulation of capital, and, finally 'labor', which[4] used the built environment as a means of consumption and for its own reproduction (Harvey, 1976a, p. 266). Relying upon Marx (1967, Volume 2, p. 210, 1973, pp. 681-7), Harvey distinguished between 'fixed capital' items (such as factories and railroads) used in production, and 'consumption fund' items (for example, houses, roads and parks) used in consumption. Harvey went on to examine the conflict arising from 'labor's use of the consumption fund' (Harvey, 1976a, p. 266).

Noting that the domination of capital over labour was basic to the capitalist mode of production, Harvey pointed out that industrial capitalism had forced a separation between labour's place of work and place of residence or living. Harvey was mainly concerned with the conflict arising in the place of living where capital attempted to manipulate labour's consumption so that it would fall into line with the maintenance of the accumulation of capital (Harvey, 1976a, pp. 266-7). 'Capital in general' could not afford the outcome of struggles around the built environment to be determined solely by the relative powers of labour, the appropriators of rent and the construction faction. Thus 'capital in general' often intervened through the agency of state power (Harvey, 1976a, p. 271). Such intervention might appear to be favourable to labour although this was largely illusory. Harvey demonstrated, for example, how homeownership, facilitated by government, provided housing for labour but, at the same time, created a debt-encumbered and therefore socially stable working class. Homeownership promoted working-class allegiance to the principle of private property and the ethic of possessive individualism, thereby providing stimulus for continued capital accumulation, and also fragmenting the working class into 'housing classes' of homeowners and tenants (Harvey, 1976a, pp. 272-3). In so far as capitalism had survived, Harvey believed that capital had dominated labour in the living place, largely by defining both the standard of living and the quality of life, 'in part through the creation of built environments that conform to the requirements of accumulation and commodity production' (Harvey, 1976a, p. 279).

In another case study, drawing upon Walker's doctoral research at Johns Hopkins University (Walker, 1977), Harvey explored the estrangement of labour from nature, viewing the various planning strategies to 'bring back nature into the city' as merely capitalist attempts to compensate for the labourers' alienation from the product of their labour (Harvey, 1976a, p. 285). 'Nature' was thus defined as a leisure activity to be consumed rather than as the object of people's labour. Furthermore,

> Bourgeois art, literature, urban design, and "designs for urban living", offer certain conditions in the living place as compensation for what can never truly be compensated for in the work place . . . And if labour refuses to be drawn in spite of all manner of seductions, blandishments, and a dominant ideology mobilized by the bourgeoisie, then capital must impose it because the landscape of capitalist society must in the final analysis respond to the accumulation needs of capital, rather than to the very real human requirements of labor (Harvey, 1976a, p. 288).

Harvey believed that conflicts in the living place of labour were 'mere reflections of the underlying tension between capital and labor' (Harvey, 1976a, p. 289), which were mediated by the appropriators of rent and the construction faction of capital. But when problems of overaccumulation arose, the mechanisms for 'mystification' began to crumble and labour recognised that 'the bargain that it had struck with capital is no bargain at all but founded on an idealized mystification' (Harvey, 1976a, p. 290).

Harvey went on to describe three 'general situations' with regard to class struggle around the built environment, which he referred to as 'points on a continuum of possibilities' (Harvey, 1976a, pp. 290, 293). In the first situation, 'possessive individualism', each worker competed with others for a living place located away from employment centres, giving rise to the city form described by neo-classical urban land-use models. In the second situation, 'parochial community action', communities acted in their own interest, giving rise to an island-like city form. The third situation was that of a 'class conscious proletariat' actively struggling against exploitation in both the work place and the living place. Here, rent was not the result of competitive bidding or of community action but, according to Harvey, was clearly forced upon labour in the course of its class struggle against appropriators (Harvey, 1976a, p. 293).

Harvey concluded 'Labor, Capital, and Class Struggle around the Built Environment in Advanced Capitalist Societies' by noting that in the 'deep underlying conflict between capital and labor' (Harvey, 1976a, p. 294) was to be found the unity between work-based and community-based conflicts. These two types of conflict

are not mere mirror images of each other, but distorted representations, mediated by many intervening forces and circumstances, which mystify and render opaque the fundamental underlying class antagonism upon which the capitalist mode of production is founded. And it is, of course, the task of science to render clear through analysis what is mystified and opaque in daily life (Harvey, 1976a, p. 295).

In this article, Harvey placed his previous research on the struggle over housing (Harvey and Chatterjee, 1974; Harvey, 1974b, 1975a, 1975c) within the broader context of class struggle. He identified the struggle as being between subjective classes yet based upon struggle between objective classes, to use one set of terms he had previously developed (Harvey, 1974b). From the point of view of another paper (Harvey, 1975a), the article dealt with a secondary force of class structuration which gave rise to housing consumption classes. It was the drive to further the accumulation of capital in a capitalist society which ultimately led to the development of these consumption classes.

In his second article published in 1976, Harvey turned his attention to Marx's theory of the State (Harvey, 1976b). The immediate context in which 'The Marxian Theory of the State' appeared was a debate on national economic planning in *Antipode*, initiated by Wolf (1976) and continued by Edelson and Eliot Hurst (1976) and Harvey. Harvey began by apologising for the abstract nature of the article.

> Obviously, the theory remains a mere abstraction until it is put to
> work. All I can say is that the theoretical statement which follows
> has been helpful to me in my studies of the urbanization process in
> Britain and the United States and that I have also found it helpful as
> a means to think about the prospects for State action in the present
> state of capitalist development (Harvey, 1976b, p. 81).

Harvey's main aim was to show that the State in capitalist society must,
of necessity, perform certain basic minimum tasks in support of the
capitalist mode of production. In doing so, he drew upon the work
of Marxists such as Miliband (1969), Poulantzas (1973, 1975, 1976),
O'Connor (1973) and Laclau (1975).

According to Harvey, Marx viewed the State as arising out of society
yet placing itself above, and increasingly alienating itself from, society.
Its instruments of domination, especially the law, the power to tax and
the power to coerce, could be transformed into instruments of class
domination. Here, the ruling class had to exercise State power in its own
class interest at the same time as it argued that State actions were for
the good of all. Harvey pointed out that the ruling class often achieved
this by using ideology to represent its own interests as the universal
interests of society (Harvey, 1976b, pp. 81-2).

The 'system of bourgeois law' was based upon notions such as freedom, right and equality, which were imbued with meanings determined
by the exchange relations of capitalism (Harvey, 1976b, p. 83). This
system of law thus attempted to preserve the equality and freedom
of exchange, protect property rights and enforce contracts, preserve
the mobility of labour and capital, regulate the '"anarchistic" and destructive' aspects of capitalist competition and arbitrate between the
different factions of capital (Harvey, 1976b, p. 84). The State played an
important role in providing public goods and social and physical infra-
structures necessary to capitalist production but which no individual
capitalist would find it possible to provide at a profit. The State also,
inevitably, according to Harvey, became an economic crisis manager on
behalf of capitalism (Harvey, 1976b, p. 84).

Harvey then departed from 'deriving a theory of the capitalist State'
from Marx's writings in order to consider two general points about
'bourgeois social democracy' (Harvey, 1976b, p. 84). First, bourgeois
social democracy separated economic and political power, the basis of
the former being private property rights, of the latter being universal

suffrage. Thus it embodied a strong ideological and legal defence of equality, mobility and freedom at the same time as being 'highly protective' of the basic economic relationship between capital and labour, although 'wealth has to employ its power indirectly' in the political sphere (Harvey, 1976b, pp. 84-5). Secondly, as Gramsci (1971) had elucidated, in order to preserve its political hegemony, the ruling class often had to make concessions not in its own immediate economic interests. Drawing upon his previous work (Harvey, 1976a), Harvey gave the example of how state policies supporting homeownership provided minimum living standards for the working class but also opened up a new market for capitalist production and encouraged support for the principle of private property which was so fundamental to capitalism.

> Under these conditions, the relationships between the State and the class struggle become somewhat ambiguous; it is inappropriate, therefore, to regard the capitalist State as nothing more than a vast capitalist conspiracy for the exploitation of workers (Harvey, 1976b, p. 85).

Harvey also believed that the conception of the State as a superstructural form was appropriate for purposes of theoretical analysis but was singularly inappropriate if 'naively projected into the study of the history of capitalist societies' (Harvey, 1976b, p. 87).

> The bourgeois State did not arise as some automatic reflection of the growth of capitalist social relations. State institutions had to be painfully constructed and at each step along the way power could be and was exercised through them to help create the very relations which state institutions were ultimately to reflect. Marx plainly did not regard the State as a passive element in history (Harvey, 1976b, p. 87).

Marx believed that the economic basis and superstructure of society came into being simultaneously and dialectically, not sequentially (Harvey, 1976b, p. 88).

In concluding 'The Marxian Theory of the State', Harvey posed three questions 'which will likely be resolved as much through concrete material investigations of history as through further theoretical analysis' (Harvey, 1976b, p. 89). The questions involved the extent to which the various instrumentalities of state power were autonomous with respect to the path of capitalist development, the degree to which the capitalist

State might vary in its form and structure, and which structures and functions of the State were 'organic', or necessary, to the capitalist mode of production (Harvey, 1976b, p. 89).

In a paper published in 1978, Harvey examined one of the instrumentalities of the capitalist State. Entitled 'On Planning the Ideology of Planning', the paper was originally presented at a conference on planning theory in the 1980s and was published together with other conference papers (Burchell and Sternlieb, 1978). Adapting his vocabulary and approach so as to communicate clearly with those not familiar with Marxist theory, Harvey presented a view of the planner in the context of what he referred to as 'a sociological description of society which sees class relations as fundamental' (Harvey, 1978a, p. 215). Noting that it was the task of the planner to intervene in the production of the built environment to ensure its proper management and maintenance, Harvey asked, 'useful or better for what and to whom?' (Harvey, 1978a, p. 213).

Harvey then explained to his listeners and, subsequently, his readers, the way in which the built environment was used for different purposes by different classes; how urban planning was an instrument of the capitalist State to overcome crises, reproduce capitalist social relations and maintain the accumulation of capital through the management of the built environment; how the planning ideologies based upon the notions of social justice and the beautiful city in the 1960s were fundamentally attempts to compensate in the sphere of consumption for the alienation the working class experienced in the sphere of production; and how, in the 1970s, in order to cope with emerging underconsumption crises, the planner's ideology had to change.

> The pursuit of the city beautiful is replaced by the search for the city efficient, the cry of social justice is replaced by the slogan "efficiency in government" and those planners armed with a ruthless cost-benefit calculus, a rational and technocratic commitment to efficiency for efficiency's sake, come into their own (Harvey, 1978a, p. 228).

But since ideologies do not easily change, 'it becomes necessary to plan the ideology of planning' (Harvey, 1978a, p. 229). Harvey invited planners to 'step aside and reflect upon the tortuous twists and turns in our history' and to recognise the mystification which lay in the presupposition of 'harmony at the stillpoint of the turning capitalist world' (Harvey, 1978a, p. 231). He concluded:

And we might even come to see that it is the commitment to an alien ideology which chains our thought and understanding in order to legitimate a social practice that preserves, in a deep sense, the domination of capital over labor . . . We might even begin to plan the reconstruction of society, instead of merely planning the ideology of planning (Harvey, 1978a, p. 231).

In this paper, Harvey used his research as a basis for the 'demystification' of urban and regional planning. He was attempting to raise the consciousness of planners with regard to the role they played in reproducing the capitalist social order.

'Government Policies, Financial Institutions and Neighbourhood Change in U.S. Cities' (Harvey, 1977a), which drew heavily upon 'The Political Economy of Urbanization in Advanced Capitalist Countries: the Case of the United States' (Harvey, 1975c), represents Harvey's latest synthesis of research on the Baltimore housing market within the context of the overriding capitalist concern for the accumulation of capital. In contrast, in an article entitled 'The Urban Process under Capitalism: a Framework for Analysis', Harvey presented the framework which lay behind his theoretical and conceptual analyses of urbanisation under advanced capitalism (Harvey, 1978b). Published in 1978 in the *International Journal of Urban and Regional Research*, it had been presented originally to a workshop on 'Marxism, imperialism and the spatial analysis of peripheral capitalism' in Amsterdam in May 1977 (Walker, 1978, p. 36). Harvey noted that the article was 'a distillation out of a larger and much vaster work (which may see the light of day shortly)' and that the framework had emerged 'as the end-product of study' and was not an arbitrary imposition upon his work from the beginning (Harvey, 1978b, p. 130).

Harvey understood the urban process under capitalism in relation to the twin themes of accumulation and class struggle. Accumulation was the means whereby the capitalist class reproduced both itself and its domination of labour. Thus the two themes were but 'different sides of the same coin — different windows from which to view the totality of capitalist activity' (Harvey, 1978b, pp. 101-2). Harvey went on to examine the laws of accumulation by sketching the structure of the flows of capital within the capitalist system of production and realisation of value. He outlined the three circuits of capital, which were, in effect, 'a summary of Marx's argument in *Capital*' (Harvey, 1978b, p. 104).

The primary circuit of capital involved the production of value and surplus value by labour, which sold its labour power in exchange for wages, which were then used by labour for their own reproduction and to purchase luxury goods (Harvey, 1978b, p. 104). Harvey pointed out that a contradiction between the action of individual capitalists and their aggregative results led to a tendency towards overaccumulation in the primary circuit. 'Too much capital is produced in aggregate relative to the opportunities to employ that capital' (Harvey, 1978b, p. 106).

Harvey presented the secondary circuit of capital in relation to the formation of fixed capital and the consumption fund. Surplus capital from the primary circuit was invested in producer and consumer durables and in the physical framework for production and consumption, that is, the built environment (Harvey, 1978b, p. 106). Because of the high cost of built environment items, and thus the tendency for individual capitalists to undersupply the collective needs for production, a capital market was necessary to co-ordinate investment in the built environment on behalf of the capitalist class (Harvey, 1978b, pp. 106-7). The flow of capital into the built environment specified by the secondary circuit is the context in which Harvey's research on urbanisation and housing markets has to be placed.

The tertiary circuit of capital comprised investment in science and technology, in order to revolutionise the productive forces in society, and in social expenditures relating primarily to the reproduction of labour power. The flow of capital into the development of ideological, military and other means for the co-optation, integration and repression of the labour force was also part of the tertiary circuit (Harvey, 1978b, p. 108).

Harvey used a diagram to portray the overall structure of relations comprising the circulation of capital among the three circuits (Harvey, 1978b, p. 109, Figure 3). He admitted that the diagram looked very 'structuralist-functionalist' because of the method of presentation, but he could conceive of no other way 'to communicate clearly the various dimensions of capital flow' (Harvey, 1978b, p. 108). Harvey believed that in so far as the three circuits portrayed 'capital movements into the built environment (for both production and consumption) and the laying out of social expenditures for the reproduction of labour power', they provided 'the structural links we need to understand the urban process under capitalism' (Harvey, 1978b, p. 114).

The acid test of any set of theoretical propositions comes when we seek to relate them to the experience of history and to the practice of politics. In a short paper of this kind I cannot hope to demonstrate the relations between the theory of accumulation and its contradictions on the one hand, and the urban process on the other in the kind of detail which would be convincing. I shall therefore confine myself to illustrating some of the more important themes which can be identified (Harvey, 1978b, p. 115).

Harvey noted that the theoretical framework he had outlined indicated that the impulses deriving from the tendencies to overaccumulate and to underinvest should be rhythmic rather than constant. He illustrated that the historical evidence was consistent with rhythmic patterns of investment and of movement of capital and labour (Harvey, 1978b, pp. 115-20).

While I am not attempting in any strict sense to "verify" the theory by appeal to the historical record, the latter most certainly is not incompatible with the broad outlines of the theory we have sketched in (Harvey, 1978b, p. 121).

Other historical evidence, such as that relating to devaluation, was also viewed as being generally consistent with Harvey's theoretical argument.

Harvey then introduced the dimension of class struggle, emphasising that the central point of tension between capital and labour was to be found in the work process. But class struggle also took place in the sphere of consumption. Harvey called this 'displaced class struggle' (Harvey, 1978b, p. 125) which, elsewhere, he had designated as 'subjective class struggle' (Harvey, 1974b, p. 251). Relying heavily upon 'Labor, Capital, and Class Struggle around the Built Environment in Advanced Capitalist Societies' (Harvey, 1976a), Harvey concluded by considering various facets of activity within the housing market as examples of displaced class struggle (Harvey, 1978b, pp. 126-30).

### Pluralism in Geography and Academic Freedom

As noted in Chapter 3, in 1972, Berry, a leading urban geographer in the spatial science tradition, had criticised the idealism in Harvey's 'Revolutionary and Counter-revolutionary Theory in Geography and the Problem of Ghetto Formation', while Harvey, in reply, defended his position (Berry, 1972b; Harvey, 1972e). This dialogue was renewed in 1974 when Berry reviewed *Social Justice and the City* and in 1975 when Harvey reviewed Berry's *The Human Consequences of Urbanisation*.

Berry's review of *Social Justice and the City* was published in

*Antipode* (Berry, 1974a). He observed that Harvey was 'stimulating, provocative, and even on occasion profound . . . although he may sit a long time waiting for the revolution' (Berry, 1974a, pp. 143-4). Berry went on to argue that the control of society was no longer primarily economic, as Marx had indicated, but political, and that 'while Harvey is waiting for the transformations that will produce more just forms of industrial capitalism, the actual transformations will have produced the post-industrial world' (Berry, 1974a, p. 145).

In his reply to the review, Harvey accused Berry of producing a 'formula review' which, although it managed accurately to précis parts of *Social Justice and the City* went on to criticise only by 'innuendo' (Harvey and Berry, 1974, pp. 145-6). Harvey defended Marx's views and disagreed with Berry that a more just society would be the result of emerging political trends. He described Berry's own work as 'highly professional and technocratic in its approach to subject matter but highly pragmatic when it comes to power' (Harvey and Berry, 1974, p. 147). Berry was given a chance to reply to Harvey's remarks and he accused Harvey of a 'formula response' in which he impugned the motives of the critic. Unlike Harvey, Berry believed that change could be produced from within 'the system . . . The choice, after all, is not that hard: between pragmatic pursuit of what is attainable and revolutionary romanticism, between realism and the heady perfumes of flower-power' (Harvey and Berry, 1974, p. 148). Harvey retorted that Berry was obviously not interested in the important issues of urban theory that had been raised in *Social Justice and the City*.

> Indeed he has constructed a most beautiful "Catch-22" formula for dealing with all "radical" literature. In a recent review of a book by Eliot-Hurst in the *Geographical Review* he argues that the "radicals" cannot command any credibility until they produce a real content and in his review of *Social Justice* he argues that the content is not worthy of consideration because it is based on "radical" assumptions (whatever they may be) (Harvey and Berry, 1974, p. 149).

Berry was allowed the last word. He ended the 'discussion' with three terse sentences. 'Once again, Harvey chooses to obfuscate. I only reviewed Hurst in the *Review*, and Harvey here. Two individuals a Catch-22 do not make' (Harvey and Berry, 1974, p. 149).

In his review of Berry's (1973) *The Human Consequences of Urbanisation* in the *Annals of the Association of American Geographers*, Harvey referred to the book as 'all fanfare and no substance' (Harvey,

1975e, p. 99). He believed that a critical reading of it revealed both the bankruptcy of typical Anglo-American urban theory and that 'it is scholarship of the Brian Berry sort which typically produces such messes' (Harvey, 1975e, p. 99). Berry did not recognise that 'in a fundamental sense urbanization is economic growth and capital accumulation' (Harvey, 1975e, p. 102). Thus, although he had done a great deal to put professional geography 'onto the map', Berry's book revealed that he had nothing of substance to say about a subject on which he was supposed to be an expert (Harvey, 1975e, p. 102). Harvey believed that it was analyses like Berry's, with an attachment to the 'liberal virtue of objectivity', a faith in 'technocratic "scientific" solutions' and a naive optimism, which had led America into the Vietnam War.

> A critical reading of Berry's *The Human Consequences of Urbanisation* must, for this reason, give us pause, for Brian Berry is influential and important. If *The Human Consequences of Urbanisation* provides us with a yardstick to judge, his influence is potentially devastating (Harvey, 1975e, p. 103).

Berry did not respond in publication to Harvey's review although, in an interview published in the *Journal of Urban History* in 1978, he pointed out that, in *Social Justice and the City*, Harvey started with the premiss that a capitalist city was inherently unjust. In *The Human Consequences of Urbanisation*, Berry had dealt 'much more with the nature of things as they are' (Halvorson and Stave, 1978, p. 222), advocating the use of planning within the capitalist system to ameliorate present problems and avoid future ones.

> I did not write a political tract. Harvey did . . . The Union of Socialist Geographers has decided that I'm one of several people who apparently have to be destroyed. And that's fine if they want to battle in the political arena . . . And there's no more amusing thing than goading a series of malcontents and kooks and freaks and dropouts and so on, which is after all what that group mainly consists of. There are very few scholars in the group (Halvorson and Stave, 1978, p. 223).

It is not clear whether Berry regarded Harvey as one of the 'very few scholars' in the Union of Socialist Geographers or not. In a recent review of Gale and Olsson's (1979) *Philosophy in Geography*, he commended in particular the essays by Tuan and Harvey ('Population, Resources, and the Ideology of Science' – Harvey, 1974d) 'for their clarity of exposition' (Berry, 1980a, p. 197). Later in the review, Berry noted that the authors of the essays in the book included 'dialecticians ranging in beliefs from Marx's crude economic materialism to the contemporary Parisian view of dialectics as a pure art form' (Berry, 1980a, p. 197). Whether Berry viewed Harvey as presenting a 'crude economic materialism' or not is unclear. However, in his Presidential Address to the Association of American Geographers in April 1980, Berry again referred favourably to Harvey's (1974d) essay. He implied that it presented a Marxist perspective 'stripped of the crude nineteenth century economism of historical materialism and the equally crude two-class polarization of capitalists and workers' and viewed society as able to create its own history, transforming the environments with which its members interacted (Berry, 1980b, p. 457). Berry has yet to publish comments on Harvey's post-1974 Marxist theory of urbanisation in advanced capitalist countries.

This 'often virulent' (Johnston, 1979, p. 171) exchange between Harvey and Berry illustrates the difficulty of genuine dialogue between those holding different philosophical and associated ideological positions in geography. Personal antagonisms and hidden antipathies tend to confuse the issues involved and obstruct understanding. However, the philosophical and ideological gulf between Harvey and Berry was substantial. On the one hand, Harvey's Marxist geography was seen by Berry to incorporate unrealistic and subjectively based ideological beliefs inadmissible to a scientific understanding of urban reality. On the other hand, Berry's liberal technocratic geography was viewed by Harvey as suffused with apologetics for an unjust *status quo*, using capitalist means which could only result in capitalist ends.

Another example of an ideological debate complicated by 'personal' factors was a clash, in print, between Harvey and G.F. Carter. In 1977, in a communication in the *Professional Geographer* entitled 'A Geographical Society should be a Geographical Society', Carter noted with dismay the appearance of organisations representing Marxist and 'gay' geographers at meetings of the Association of American Geographers (Carter, 1977, p. 101). He pointed to a stream of 'disruptive, contentious, politically loaded resolutions' put forward by Marxist geographers, particularly one by Harvey and Peet condemning 'Pinochet, Chile, the CIA, and the USA', which threatened to 'tear the Association to pieces' (Carter, 1977, pp. 101-2). Carter advocated that the Association stay out of political and moral controversies and confine itself to professional business, to 'geography by geographers and for geographers'

(Carter, 1977, p. 102). Harvey, in response, defended his criticism of Pinochet but also encouraged 'political, ideological and scientific pluralism' in geography. 'And it is precisely to open up the Association of American Geographers to such a prospect that I insist upon putting those "contentious" resolutions' (Harvey, 1977c, p. 407). Harvey then accused Carter of 'blind cold-war politics' in the McCarthy Era when Carter was apparently responsible for 'hounding out' Marxist and Maoist scholars.

> Professor Carter divided and nearly destroyed a department (ironically, the very one in which I am now Professor) and split a whole university community into warring factions, as he pursued his holy war against "contentious and divisive" thought (Harvey, 1977c, p. 407).

Such activities, wrote Harvey, imposed a homogeneity of thought and practice within geography which had nothing to do with scholarship. He concluded that it was Carter's ideas, not those of Marxist geographers, which were not in the interest of the Association of American Geographers (Harvey, 1977c, p. 407). In this encounter, Harvey demonstrated once again that a neutral ideological stance was often merely a guise for a commitment to conservatism and the upholding of the *status quo*.

In 1978, Harvey was invited to take part in a dialogue between 'urban sociologists in the Chicago School tradition and Marxian challengers to that tradition' (Harvey, 1978c, p. 28). The result was an article entitled 'On Countering the Marxian Myth – Chicago Style', published in *Comparative Urban Research* (Harvey, 1978c). Harvey argued that the 'struggle' between the two traditions was a struggle 'to establish a hegemonic system of concepts, categories and relationships for understanding the world, . . . a struggle for language and meaning itself' (Harvey, 1978c, p. 28). It was therefore difficult for Chicago School sociologists to understand Marxism unless they paid close attention to the integrity of Marxian concepts.

> I can testify to some of the extraordinarily complex problems which arise when confronting the Marxian meanings from out of a "bourgeois" positivist and analytical tradition. It has taken me almost seven years to acquire even a limited fluency in the use of Marxian concepts, and I am always being surprised by the continued unfolding of new patterns of meanings and relationships (Harvey, 1978c, p. 29).

'Bourgeois social science' took an analytic and empirical stance which invariably led to an excessive fragmentation of knowledge, although crises in Western capitalism presented problems which demanded a more systematic and synthetic approach. A theory which dealt directly with crisis formation, as did the Marxian theory, then began to be taken seriously by both the population at large and academics. A capitalist crisis thus generated a resurgence of interest in Marxian theory to the same degree that it provoked a crisis within bourgeois social science (Harvey, 1978c, p. 33).

Harvey characterised the Marxian approach as 'wholistic'. It worked with a sense of totality as a set of internally related parts, the view presented in the final chapter of *Social Justice and the City*, although Harvey did not use the term 'operational structuralism' in the article (Harvey, 1978c, p. 33). He pointed out that bourgeois sociologists often misinterpreted the relational and dialectical method of Marx as an attempt to import an economic viewpoint into sociology. But bourgeois economists often accused Marxists of seeking to introduce a sociological component into economic analysis. This was partly because 'the Marxian challenge . . . attempts to be subversive of all disciplinary boundaries'. Furthermore,

> Marx did not disaggregate the world into "economic", "sociological", "political", "psychological", and other factors. He sought to construct an approach to the totality of relations within capitalist society . . . There are plenty of controversies within the Marxian tradition . . . There are various schools of thought (including one which is very "economistic" and "reductionist") (Harvey, 1978c, p. 35).

In order to establish the nature of the Marxian theory as he saw it, Harvey decided 'to illustrate, very briefly, the thinking which underlies my current academic practice as I struggle to grasp the real meaning of the "urban question"' (Harvey, 1978c, p. 36). There followed an analysis of urbanisation taken mainly from three previous articles (Harvey, 1975d, 1976a, 1978b), constructed around a consideration of the twin themes of accumulation (Harvey, 1978c, pp. 36-40) and class struggle (Harvey, 1978c, pp. 40-3). Harvey acknowledged that Marx's theory of accumulation 'sounds (and reads) rather "economistic"' (Harvey, 1978c, p. 36) and that the analysis of urbanisation in relation to capital accumulation as he presented it 'probably sounds very economistic and reductionist' to 'bourgeois ears' (Harvey, 1978c, p. 39). But, Harvey argued,

Bourgeois social science typically engages in reifications, invariably representing social relations as things; . . . the social relations between capitalists, laborers, and landlords are reduced . . . to relations between the "factors of production", land, labor and capital . . . The whole thrust of the Marxian argument is, of course, to concentrate on the social meaning of things. Starting with the physical artifact that is the city, we can reach out, step by step, into the myriad social relations (between landlords and financiers, building laborers, artisans and capitalist builders, between users and producers, between the state and individuals, between communities and speculators, etc.) and into the extraordinary complexity of interactions, conflicts, coalitions within a framework of institutional arrangements, all of which lead to the creation of this physical landscape (Harvey, 1978c, pp. 39-40).

Harvey concluded that the 'bourgeois academic' could not understand Marxian thought without ultimately practising it. To construct knowledge required an active involvement in the processes of social change. In struggling to change the world, 'we change both the world and ourselves' (Harvey, 1978c, p. 44). The only sure path to that knowledge which has the capacity to change the world was to engage in such a struggle. Bourgeois academics would have to cease to be bourgeois and go to 'the other side of the barricades' if they were really to understand what 'the view from inside, the view from the standpoint of labor, is really all about' (Harvey, 1978c, p. 44).

In 'On Countering the Marxian Myth — Chicago Style', Harvey used both his research on the Baltimore housing market and the Marxian analytical framework to attempt to 'raise the consciousness' of those holding an opposing theoretical viewpoint. While genuinely encouraging dialogue, and willingly entering into it, Harvey at the same time had no illusions that the 'struggle for language and meaning itself' (Harvey, 1978c, p. 28) would be easily won.

In two short newspaper articles published in the *Baltimore Sun* in 1978, Harvey defended the principle of academic freedom. In the first, entitled 'The Subversion of Tenure on the Homewood Campus', Harvey sought to defend academic freedom at Johns Hopkins from an attack by its Board of Trustees (Harvey, 1978d). The Board had initially affirmed its commitment to academic freedom but had then decided that

a grant of tenure confers no right to permanent or continued employment where the position to which the tenured faculty member has been appointed is abolished, whether the abolition of the position is attributable to redirection of the university's interests, curricular needs or financial exigency (Harvey, 1978d).

Harvey pointed out that it was hard enough to protect the positions of 'genuine free thinkers . . . [and] the dissidents in academia' with the tenure principle intact. The Board of Trustees had also ruled that in case of conflict between its decision and any faculty promulgation, its own statement would prevail. To Harvey, this meant the end of any remnants of academic freedom and faculty self-governance at Johns Hopkins. But it also meant that academics were made aware that they served 'a bunch of bankers, lawyers and corporate heads . . . rather than . . . the American public in general'. The trustees therefore 'may well have put faculty unionization on the agenda. Even at the Hopkins' (Harvey, 1978d).

The second article was entitled 'Karl Marx and the Boundaries of Academic Freedom' (Harvey, 1978e). It was published in a slightly revised version later in 1978 in *The Progressive* (Harvey, 1978f). In it, Harvey recounted an experience he had in 1969 just after he had arrived in the United States from England. Participating in a call-in radio talk show, he had admitted that the British had done things they were ashamed of during the United States War of Independence but that citizens of the United States should also be ashamed of what was then happening in Vietnam.

> The response was electric . . . I was variously advised that I had no rights as a foreigner to say anything (although I did carry a draft card in my pocket), that I should go back where I belonged . . . and that I was a very good reason why no one should ever send his son to the Johns Hopkins (Harvey, 1978e).

Harvey found this response perplexing as he had been raised in 'that liberal intellectual tradition in which reasoned consistency is highly valued'. He began to reflect upon the innumerable contradictions which pervaded American thought and practice, in both its foreign and domestic policies. He discovered, for instance, that the preamble to every piece of Congressional housing legislation since 1945 asserted that every citizen had a right to a decent house in a suitable living environment. However, his experience of housing conditions in part of Baltimore contradicted such ideals. Shortly after his talk-show experience, Harvey reported, he began to read the writings of Karl Marx.

> He placed great emphasis upon the "contradictions of capitalism" and that caught my attention. I read on, intrigued. And the more I read, the more it seemed to make sense. At the bottom of it all, Marx argued, lies a tension between capital and labor . . . We could, Marx suggested, create a theory to explain the contradictions at the same time as it would help us to overcome them. I found this very exciting and started to work at it. And lo and behold, one by one, the contradictions which had so perplexed me crumbled before the power of the analysis (Harvey, 1978e).

Harvey explained that he was also forced to accept the social consequences of Marx's perspective. Once this was made apparent in his writings, Harvey found that many of his academic colleagues began to look upon him with suspicion. He referred to Berry's comment about Marxist geographers being 'kooks, freaks, dropouts and malcontents' (Halvorson and Stave, 1978, p. 223) but protested that he had never felt saner in his life. Job offers suddenly dropped off. 'What I began to discover through experience was the meaning of "repressive tolerance"' (Harvey, 1978e). Freedom of thought within the university was actually confined within certain 'fuzzy' boundaries. The closer academics moved towards those boundaries, the less they were tolerated and the more repression they experienced. Harvey referred to the furore over Bertell Ollman's appointment at the University of Maryland. It was feared that Ollman ('whose whole record bespeaks a desire for diversity') would not be open to alternative ideas, a judgement which overlooked the fact that 'there are a whole raft of other departments which have systematically excluded Marxists all along' (Harvey, 1978e).

> Of course, State Senator John J. Bishop, Jr., . . . gave the real game away. This [Ollman's] appointment, he said, "is as ridiculous as it would be to appoint Henry Ford to head the political department at Moscow University". In my liberal days I would have found that remark deeply perplexing because I always thought the point was that the United States was supposed to be different from the Soviet Union and there was repression there and freedom of expression here. But I now see that Senator Bishop is just trying to tell it like it really is (Harvey, 1978e).

Harvey concluded by comparing the report of a Commission on Subversive Activities to the Maryland General Assembly in 1949 with the activities of the CIA in Chile, which were approved by Kissinger and

Nixon. Harvey asserted that both the 'subversive activities' and the actions of the CIA were treasonable conspiracies against democracy. He noted that in his liberal days he would have been surprised that Kissinger was still at large. 'But I am older and wiser now. I simply recognize that it all depends, in the end, on whose side you are on. And that, of course, is what Marx meant by class struggle' (Harvey, 1978e).

### Revolutionary Class Struggle in Paris and Central America

In the next paper to be considered, Harvey took his reader on a tour through the Basilica of Sacré-Coeur as it now stands upon the Butte Montmartre in Paris, as well as through the history of Paris since the seventeenth century associated with the church and its site. The article was published in 1979 in the *Annals of the Association of American Geographers* under the title of 'Monument and Myth'. In it, Harvey departed from his analysis of the urban process under capitalism and the theoretical dimension of Marxist geography. Through a concrete historical study, he compiled an account of the variety of meanings of a landscape artifact. The Basilica symbolised one thing to the people who commissioned it but another quite different thing to those who struggled against its construction. However, the study dealt with concerns apparent in Harvey's previous work in that it presented a case study of the role of class struggle in urban history. Furthermore, Harvey had stated as early as 1974 that 'demystification' was an important task for the Marxist scholar (Harvey, 1975b, p. 55) and the article is an exemplification of this. That demystification is the task of 'Monument and Myth', published as it was in the most important and conservative of American journals of geography, is apparent from its closing passage:

> The building hides its secrets in sepulchral silence. Only the living, cognizant of this history, who understand the principles of those who struggled for and against the "embellishment" of that spot, can truly disinter the mysteries that lie entombed there and thereby rescue that rich experience from the deathly silence of the tomb and transform it into the noisy beginnings of the cradle.
>
> All history is, after all, the history of class struggle (Harvey, 1979, p. 381).

Harvey outlined the history of the Roman Catholic cult of the Sacred Heart of Jesus, founded in the seventeenth century, and its opposition to the principles of the French Revolution. He detailed the Prussian siege of Paris in 1870 to 1871 and the establishment of the Paris Commune in March 1871, leading to the outbreak of violence between Parisian conservatives and radicals. Two French generals, attempting to disarm the Paris populace after the Prussian siege had been lifted, were shot by a crowd on 18 March 1871.

> This incident is of crucial importance to our story . . . In the months and years to come, as the struggle to build the Basilica of Sacré-Coeur unfolded, frequent appeal was to be made to the need to commemorate these "martyrs of yesterday who died in order to defend and save Christian society" [De Fleury, 1903, p. 88] (Harvey, 1979, p. 370).

The French military withdrew and besieged Paris in early April 1871, eventually overcoming the city in 'one of the most vicious blood-lettings in an often bloody French history' (Harvey, 1979, p. 372). After relating the abuse and final execution, on the hillside of Montmartre, of Eugene Varlin, an 'intelligent, respected and scrupulously honest committed socialist, and brave soldier', Harvey observed that 'the left can have its martyrs too. And it is on that spot that Sacré-Coeur is built' (Harvey, 1979, p. 372). For, after the fall of the Commune, the newly-appointed Archbishop of Paris accepted the task of building a sacred monument to the defeat of the Commune and its crimes against the State and the Church (Harvey, 1979, p. 376). The first stone was laid in 1875 and the Basilica was finally consecrated in 1919. But Harvey questioned what was interred there.

> The spirit of 1789? The sins of France? The alliance between intransigent catholicism and reactionary monarchism? The blood of martyrs like [Generals] Lecomte and Clement Thomas? Or that of Eugene Varlin and the twenty thousand or so communards mercilessly slaughtered along with him? (Harvey, 1979, p. 381)

Thus the monument on the Butte Montmartre symbolised (meant) different things to different classes. But only those cognisant of its history and of the principles of the men and women who struggled for and against its establishment could transform 'that rich experience from the deathly silence of the tomb . . . into the noisy beginnings of the cradle' of revolution (Harvey, 1979, p. 381).

In 1980, Harvey co-authored two short articles in *The Progressive*, a popular left-wing United States monthly magazine (Koeppel and Harvey, 1980a, 1980b). Harvey and Barbara Koeppel, a contributing editor of *The Progressive*, had visited Central America early in 1980. The first article was entitled 'Nicaragua Rebuilds'. In it, the authors outlined the massive economic, social and political problems faced by the successful Sandinista Liberation Front (FSLN) and its Government of National Reconstruction after the recent civil war. The FSLN had popular support in the country, its revolution being based upon the ideologies of the theology of liberation, revolutionary Marxism and nationalism. The revolutionaries promised to develop a 'just and humane kind of socialism' (Koeppel and Harvey, 1980a, p. 42).

> Rents and mortgages on low- and middle-income housing have been cut in half. The lowest-paid government workers have received wage increases, while a ceiling of $1,000 a month has been put on the highest salaries (even those of the junta). Hospitals, schools, and universities, which used to charge fees, are now free. Speculation and profiteering have been curbed, and serious attempts have been made to control soaring food prices (Koeppel and Harvey, 1980a, p. 42).

But the new government had both leftist and rightist critics, the former seeing it as laden with representatives from the traditional elite, posing a barrier to the creation of a worker-peasant state, whereas the latter had revived the Conservative-Democratic Party which claimed to support the revolution but appealed strongly for the retaining of the 'traditional values' of individualism, free enterprise, private property, Church and family (Koeppel and Harvey, 1980a, p. 45).

The FSLN had set up Sandinista Defense Committees in every block in every town and rural area. These were charged with overseeing the return to law and order, cleaning up war damage, distributing emergency relief and implementing health measures. According to Koeppel and Harvey, some of these committees were functioning as models of participatory democracy. The FSLN did not plan to hold national elections for some time, however. It argued that a democracy could not be built when half the population was illiterate. A new model had emerged from the ruins left by Samoza and the civil war, a 'purely Sandinist model' that was neither communist nor capitalist.

> Some of the ingredients of that model are already evident. Strong government controls over credit and foreign commerce will be the key tools of central planners. Within the economy, traditional private enterprise will intermingle with cooperative and collective endeavours (particularly in agriculture) backed by decentralized community

control movements and mechanisms for worker and peasant partici-
pation in economic decision making (Koeppel and Harvey, 1980a,
p. 46).

In the second article in *The Progressive*, 'Tragedy in El Salvador',
Koeppel and Harvey outlined the starkly contrasting case of El Salvador.
By restricting its reporting to information released by 'official sources' in
El Salvador, the media of the United States had distorted the truth and
misrepresented the torture and murder of innocent people as defence
against terrorists and rebels (Koeppel and Harvey, 1980b, p. 44). The
authors quoted Amnesty International reports on the murders of Fabio
Castellano, a young Communist professor, Mario Zamora, a Christian
Democrat member of the government who had been critical of the
military's repressive actions, and Archbishop Oscar Romero. But,
Koeppel and Harvey alleged, most of the military's 'savagery' had been
directed against the peasants, bordering on genocide. A new junta, with
both civilian and military members, had promised reforms but had
made no progress. Repression had intensified. The Christian Democrats
had asked that military aid from the United States be halted and had
called for a pluralistic government to be instituted.

> Tragically, U.S. officials seem to have missed the lesson of Nicaragua:
> that brutal military regimes can squash popular uprisings, but only
> for so long; that no amount of military hardware and no number of
> U.S. advisers will prevent the people from toppling a vicious regime.
> Instead, U.S. aid will move the country one step closer to civil war,
> and American policy-makers will be responsible for the thousands
> of deaths that will follow (Koeppel and Harvey, 1980b, p. 45).

In 1979, the Baltimore Rent Control Campaign was set up to obtain
support for a ballot on the issue of rent control (Kirschmann, 1981,
pp. 192-3). The Campaign group was a coalition of about 50 community, labour and church organisations. The ballot was eventually held
on 6 November 1979, and rent control was supported by a majority
of 5,000 votes. However, the Lower and Appeals Courts later ruled
that Baltimore citizens did not have the right to initiate such controls
(Kirschmann, 1981, pp. 193-4). In June 1980, the matter was taken up
within the Baltimore City Council.

Harvey contributed to the Campaign with an article in the *Baltimore
Sun* in September 1979, republished in 1981 in a source book on rent
control (Harvey, 1981). He pointed out that whereas tenants used to

spend a quarter of their incomes on rent, they were now being forced to outlay between 35 and 50 per cent. Tenants were therefore 'in revolt' and rent control had become a pressing issue. In a previous issue of the *Baltimore Sun*, an economist had put forward a 'scholarly evaluation' that 'produced a litany of objections to rent control' (Harvey, 1981, p. 80). However, Harvey pointed out that the evidence of studies on the actual impact of rent control was mixed. For example, abandonment of housing occurred in both rent controlled and non-rent controlled cities. Abandonment was therefore not necessarily the result of rent control. Harvey believed that two conclusions could be drawn from a confusing array of evidence.

> First, the impact of controls varies according to social and economic conditions in particular cities — vigorous and growing communities can continue merrily on their way, while controls are no palliative for the long-run problems of tenants in declining areas. Second, the impact depends very much on the kind of controls imposed. What are known as "restrictive" controls freeze rents for prolonged periods. They tend to squeeze out landlords and create housing shortages unless the government steps in to take up the slack (a solution which some see as not at all bad). "Moderate" rent controls, on the other hand, are designed to avoid such problems. They . . . guarantee landlords a "fair" rate of return and permit rent hikes which can be justified by rising costs or improved maintenance . . . For legislation of this sort to work, however, tenants must be protected against arbitrary eviction (Harvey, 1981, pp. 81-2).

The Campaign was proposing that the determination of what was 'fair' would be decided by a tenant-landlord commission. But Harvey pointed out that even then rents were almost certain to continue to rise at a rate which would be 'exceedingly burdensome' for the poor in particular. He concluded that one of the most important issues to be faced by the proposed commission was the extent to which a part of the inflationary burden should be borne by landlords (Harvey, 1981, p. 82).

This short newspaper article is the only published indication that Harvey has taken part in a wider 'revolutionary activity' than that related to academic geography, the analysis of urbanisation or reporting upon revolution in Central America. Harvey's contribution was to point out the fallacies in another 'scholarly' viewpoint. In doing so, he supported the development of a situation where tenants could have more control over their own living conditions. However, Harvey did not clearly relate the local issue to 'the total economy' as he advocated in 1974 as being the way forward for the Union of Socialist Geographers (Harvey, 1975b, p. 58). As a form of 'revolutionary practice', it appeared to be but a small beginning. However, it is not possible to ascertain the extent to which Harvey was involved in the Campaign beyond the newspaper article and, consequently, these conclusions may be quite misleading. Only a more biographical approach to Harvey's writings can resolve this issue.

## From Operational Structuralism to a Marxist Analytical Framework

In the early 1970s, Harvey had characterised Marx's method as constructivist and operational structuralist and had explored its application to the concerns of *Social Justice and the City* and the population-resources debate. He had also conducted research into the Baltimore housing market. Then, in 'Absolute Rent and the Structuring of Space by Governmental and Financial Institutions', Harvey, with Chatterjee, presented an analysis of the structuring and transformation of the Baltimore housing market in terms of the application of Marx's method (Harvey and Chatterjee, 1974), one of the 'first applications of Marxist-geographical theory to contemporary problems' (Peet, 1977b, p. 255). This analysis formed the basis of five articles on urbanisation under capitalism, each extending the original account in various ways, first, in relation to class monopolies and the role of finance capital (Harvey, 1974b), then with regard to class structuration and residential differentiation (Harvey, 1975a), the economy of advanced capitalist nations and the post-World War II history of urbanisation (Harvey, 1975c), neighbourhood change and dislocation (Harvey, 1977a) and displaced class struggle around the built environment (Harvey, 1976a).

Harvey developed the theoretical side of his analysis of urbanisation under capitalism in two articles, the first on Marx's locational theory as derived from his theory of the accumulation of capital (Harvey, 1975d), the second on the Marxian theory of the capitalist State (Harvey, 1976b). 'The Urban Process under Capitalism: a Framework for Analysis' (Harvey, 1978b) is a distillation of the conceptual and theoretical framework which Harvey has constructed in these publications since 1974, based upon the two themes of capital accumulation and class struggle.

One of Harvey's aims has been to demystify the capitalist urbanisation process, that is, to demonstrate that the problems and conflicts within the urban environment are but reflections of deep structural problems inherent in the capitalist mode of production. In the pursuit of demystification, Harvey has utilised not only the development of a Marxist theoretical analysis of urbanisation. He has attempted to open up dialogue with Berry (for example, Harvey and Berry, 1974), urban planners (Harvey, 1978a) and urban sociologists in the Chicago School tradition (Harvey, 1978c). He has written popular articles on academic freedom (Harvey, 1978d, 1978e) and on revolution in Central America (Koeppel and Harvey, 1980a, 1980b) in which he addressed more broadly political issues. Harvey has also lent support to the Union of Socialist Geographers (for example, Harvey, 1975b) and the Baltimore Rent Control Campaign (Harvey, 1981), as well as raised what he has seen to be important political issues in the Association of American Geographers (see Carter, 1977, and Harvey, 1977c). The range of such activities suggests that Marxism, for Harvey, is not just an academic theory. It is rather a way of making sense of the world and its contradictions (Harvey, 1978e).

In Chapter 3 (p. 100), it was noted that in the early 1970s Harvey believed that a theory could become a revolutionary, counter-revolutionary or *status quo* theory depending on how it entered into human practice. The verification of a theory depended on its use in practice. On the other hand, he also suggested that it was necessary to develop a conceptual framework in urban studies based upon Marxist concepts and he illustrated in a number of papers that certain methods generate particular sets of conclusions. From this point of view, a revolutionary theory was one which used a conceptual framework based upon Marx's view of society and utilised Marx's method. In the first view of revolutionary theory, practice is primary. In the second view, theory is primary. Harvey's writings in the late 1970s tended to follow the second way as he went on to develop a Marxist theory of urbanisation under advanced capitalism. Harvey's apparently peripheral involvement in the Baltimore Rent Control Campaign is the only indication in his writings that he has pursued, beyond his academic work, any form of revolutionary practice in the city. Even then, the extent to which this practice was in any sense revolutionary is questionable.

Harvey's research on urbanisation under capitalism may be viewed as the application of Marx's method. In describing this method, Harvey did not use the term 'operational structuralism' in his publications after 1973. Instead, he usually referred to Marx's 'philosophy of internal relations' or the 'relational' character of Marx's method (Harvey and Chatterjee, 1974, p. 22; Harvey, 1975a, p. 355, 1978c, p. 33). However, in his first article on the Baltimore housing market, Marx's 'relational' method was described in operational structuralist terms, that is, it directed study to 'the processes of structuring and transformation together with the determinate structures that mediate these processes' (Harvey and Chatterjee, 1974, p. 22). Consequently, Harvey's analysis of the Baltimore housing market has been shaped by the operational structuralist method. In a 1978 article, he confirmed that he viewed Marx's method in the same way then as he had in *Social Justice and the City* (Harvey, 1978c, p. 33).

To what extent can Harvey's analyses in the late 1970s be called structuralist? Robey has pointed out that the simplest meaning of structuralism is

> that approach . . . which has as its object . . . the 'laws of solidarity', the 'reciprocal relations' of the different facts under observation, rather than considering these facts in isolation (Robey, 1973, pp. 1-2).

Harvey has consistently sought to relate each facet of urbanisation to its context within capitalism, particularly at the scale of the city in the advanced capitalist countries. He used the term 'structure' frequently in his writings, usually to describe something, such as an urban housing submarket (Figure 4.2, p. 113), which related a number of things or events such that the relationships tended to persist for some time in similar forms. These 'structural' relationships are maintained by various forces or processes of 'structuration'. In the case of the housing submarket, these forces included the ability to obtain credit or a mortgage, the operations of speculators and realtors and the ebb and flow of market forces (Figure 4.1, p. 111). A change in one or more of these forces or processes may bring about a transformation in the relationships of the things or events, that is, the 'structure'. This description of Harvey's use of the term 'structure' echoes operational structuralism and Marx's 'relational' approach.

A second sense of structuralism is what Gregory (1978, p. 76) has referred to as 'a form of enquiry which locates explanatory structures outside the domain of immediate experience'. Harvey expressed a similar viewpoint in Chapter 7 of *Social Justice and the City*.

> Structures are not "things" or "actions" and we cannot therefore establish their existence through observation. To define elements relationally means to interpret them in a way external to direct observation. The meaning of an observable action . . . is established

by discovering its relation to the wider structure of which it is a part (Harvey, 1973a, p. 290).

However, Harvey's use of 'structure' in this sense in his writings in the late 1970s is much more concrete than implied by Gregory's definition. The structure of the Baltimore housing market may itself be unobservable but it has immediate and direct reference to observable things and events. Duncan and Ley's (1982) critique of 'structural' Marxism did not criticise Harvey's structuralist analysis as we have considered it so far. Rather, the force of their criticism was directed towards the reification of supra-individual wholes such as 'capitalism' and the 'capitalist class' (Duncan and Ley, 1982, pp. 36, 39-41) and the imposition of predetermined categories upon empirical circumstances (Duncan and Ley, 1982, p. 49). There is no doubt that Harvey's use of 'capital', 'labour' and 'capitalism' often assumes that they are wholes actively operating independently of people's wills. The issue of the imposition of predetermined theoretical categories is considered later (p. 157).

Throughout his writings in the late 1970s, Harvey emphasised other aspects of operational structuralism. In 'The Political Economy of Urbanization in Advanced Capitalist Countries: the Case of the United States', for instance, he adopted the 'Marxian posture', which consisted of an analysis of society as made up of a substructure and a superstructure (Harvey, 1975c, p. 128). Thus, Harvey presented financial institutions as a superstructural element, mediating, to the built environment, the need to maintain the rate of capital accumulation, a substructural process. Such an analysis led to the view that 'structural' problems underlay the 'surface' problems of urban reality.

> The urban problems with which we are all too familiar (inner-city decay in American cities, sky-rocketing land and housing prices in European and Canadian cities . . .) are all surface manifestations of deep structural problems in such societies (Harvey, 1975c, p. 127).

Elsewhere, Harvey referred to the 'underlying logic of capitalism' which worked itself out differently in different situations (Harvey, 1975d, p. 21, 1978b, p. 103) and to conflicts in the living place of labour as 'mere reflections' of the 'underlying' tension between capital and labour (Harvey, 1976a, pp. 289, 294). Here, Harvey was referring to basic economic elements and processes in order to explain what was occurring at the superstructural level.

The article 'The Urban Process under Capitalism' is subtitled 'a

Framework for Analysis'. We might expect such a framework to be that of operational structuralism, but Harvey noted that his framework ˙emerged as the end product of study and [is] not one which has been arbitrarily imposed at the beginning' (Harvey, 1978b, p. 130). In fact, this framework included, for instance, the circulation of capital and its contradictions, the laws of accumulation, and class struggle. The framework for analysis should therefore be seen as the result of the application of operational structuralism, that is, it is these processes which structure and transform capitalist society and the elements (the city, the housing market) of that society. Operational structuralism guides the programme; this conceptual and theoretical framework is the result of the application of the method. In his later articles, therefore, Harvey emphasised more concrete facets of Marx's analysis of society, particularly the significance of capital accumulation and class struggle.

In three articles (Harvey, 1975d, 1976b, 1978b), Harvey made a distinction which has far-reaching implications throughout his work. The distinction is that between 'theoretical' and 'concrete historical' analyses. Harvey referred to Marx's theories of the capitalist mode of production and of imperialism, for example, as 'theoretical', whereas the various Marxist theories of imperialism, such as those of Lenin (1965) and Frank (1969), were termed 'concrete historical' analyses (Harvey, 1975d, pp. 16-18).

Marx's general theory tells us of the necessity to expand and intensify geographically. But it does not tell us exactly how, when and where. Looking at the intersection of these general arguments with concrete historical analyses, we will usually be able to identify the underlying logic dictated by capital accumulation at work (Harvey, 1975d, p. 18).

Theoretical analyses were abstract and general whereas historical analyses were concrete and particular (Harvey, 1976b, p. 87). The former dealt with the 'inner logic of capitalism', the latter with how, when and where this inner logic was worked out, in actual historical situations (Harvey, 1975d, pp. 14, 16). Harvey therefore viewed Marx's theory of the State as theoretical while analyses of the social democratic form of the State were not 'theoretical derivations but apply to an understanding of the actual history of the State' (Harvey, 1976b, p. 86). Here, Harvey was distinguishing between the abstraction necessary for theoretical analysis and the application of abstract theory in a specific context. He pointed out that Marx constructed a theory of accumulation for a capitalist mode of production in a 'pure state' (Harvey, 1975d, p. 17) and that Marx's analysis of the primary circuit of capital proceeded around 'a strictly circumscribed set of interactions with all other problems assumed away or held constant' (Harvey, 1978b, p. 104). Elsewhere, Harvey had defended Marx's abstractions in the following manner.

> Marx criticized the categories of bourgeois social science on the grounds that they are abstractly fashioned without reference to the 'relations which link these abstractions to the totality' (Ollman, 1973, p. 495). Marx's abstractions are of a different kind for they focus on such things as social relations (Harvey, 1975a, p. 355).

Quoting Lukacs (1968), Harvey went on to note that, for Marx, structures could be isolated from the social totality for the purposes of analysis but that the process of isolation was only a means towards understanding the whole. Bourgeois social science dealt with isolated fragments of knowledge, each of which was viewed as autonomous and an end in itself (Harvey, 1975a, p. 355, 1978c, p. 30).

Given this view of the way in which Marx used abstraction, or 'isolation', it is apparent that 'theoretical' analyses were the result of the abstraction process, whereas 'concrete historical' analyses resulted from the application of 'theoretical' analyses to a particular concrete context. The distinction between 'theoretical' and 'concrete historical' analyses, implicit in Harvey's publications since at least 1975 (Harvey, 1975a), enabled Harvey to treat Marx's theories as if they were part of a hypothetico-deductive structure. For, despite characterising Marx's method as a form of 'dialectical materialism' (Harvey, 1974d, p. 265), Harvey utilised deductive, not dialectical, reasoning in his research. He deduced, or derived, hypotheses out of Marx's theoretical formulations and then, usually, provided evidence illustrating the validity of the hypotheses. In 'Class Structure in a Capitalist Society and the Theory of Residential Differentiation', for instance, Harvey outlined Marx's theory of class formation, along with its development by Giddens (1973), out of which he fashioned four hypotheses (Harvey, 1975a, p. 362). Harvey went on to provide a 'general argument', drawing upon empirical material, in support of these hypotheses. In two later articles, Harvey also outlined Marx's theory, derived certain expectations from it and then advocated 'an intersection between the theoretical abstractions . . . and the materialist investigations of actual historical configurations' (Harvey, 1975d, p. 20, 1976b, p. 88). In 'The Urban Process under

Capitalism: a Framework for Analysis', Harvey sought to demonstrate the compatibility of his theory with the historical record (Harvey, 1978b, pp. 115, 121).

As well as deduction, inductive inference also played a role in Harvey's analysis of urbanisation under capitalism. Throughout his work he generalised from his analysis in Baltimore to all cities in advanced capitalist countries. For example, after examining the relationship between the structure of the housing market and class-monopoly rents in Baltimore, Harvey noted that

> These conclusions are, of course, geographically and institutionally specific to Baltimore and the United States. But a cursory examination of the literature suggests that they may be generalized to all advanced capitalist countries . . . Whether or not this is the case must be proved by future research. It seems likely, however, that the processes are general but that the manifestations are particular because the institutional, geographical, cultural and historical situations vary a great deal from place to place. In other words, the processes are general, but the circumstances are unique to each case and so, consequently, are the results (Harvey, 1974b, p. 250).

Harvey did not extend his own case studies beyond Baltimore. It could be argued that the inductive problem, discussed in *Explanation in Geography* (Harvey, 1969a, p. 37 – see p. 39 of Chapter 2 above), was simply ignored by Harvey in his analyses in the late 1970s. However, it is apparent from the above quotation that he utilised an ontological assumption that the 'logic' of capitalism was common to all cities in advanced capitalist countries. On that basis, the inductive problem had no relevance within the specified limits.

In considering the question of the imposition of predetermined categories upon empirical circumstances, it is noteworthy that Harvey's distinction between 'theoretical' and 'concrete historical' analyses put Marxist theory largely beyond criticism and falsification. The main inference to be taken from the distinction is that there is a universal structure to capitalism, an 'inner logic' (Harvey, 1975d, p. 14), which may be discerned in a wide variety of contexts. Marx's failure to predict correctly what was to occur in India after the introduction of modern machinery was viewed by Harvey as 'a failure of historical analysis, not of theory' (Harvey, 1975d, p. 16). Marx's theory was placed beyond criticism. Harvey's concrete historical analyses attempted to demonstrate 'where, when and how' (Harvey, 1975d, p. 16) the inner logic of capitalism worked itself out, rather than to verify or falsify the Marxist theoretical framework. Thus Marx's theory guided Harvey's research. Harvey's research was not designed to test the validity of the theory.

The distinction therefore becomes, or (perhaps more accurately) reflects, a separation. The synthesis, or intersection, between the theoretical and historical then becomes a serious problem. Harvey's writings have largely been theoretically focused with only secondary reference to historical and contemporary urban material, often as illustration. There are perhaps two exceptions. The first is 'Monument and Myth' (Harvey, 1979), which is a concrete historical study in which Harvey made no connection with his theoretical work, except at the end of the article where he briefly referred to class struggle. The synthesis remains problematical. The second exception is his work on Baltimore. Here, he examined the way in which communities were disrupted and destroyed through the activities of developers, speculators and financial institutions (Harvey, 1975c, pp. 140-56, 1977a, pp. 300-12, 1977b, pp. 131-7). However, even in this work, Harvey admitted that the essential theoretical proposition that the financial superstructure coordinates urbanisation with capital accumulation is treated as an assumption rather than a statement to be proved or disproved (Harvey, 1975c, p. 141). Duncan and Ley (1982, pp. 48-52) have argued that Harvey has not been very successful at a synthesis between theory and empirical research. Our study tends to support that view but a final judgement must rest upon a consideration of a greater range of material in urban, philosophical and other studies than that considered in this chapter. However, it is perhaps appropriate to suggest that the resolution of the integration between the theoretical and concrete historical relies ultimately upon the resolution of the freedom/necessity tension which exists at a fundamental and presuppositional level in Marxist thought. Theoretical research identifies the realm of necessity, the inner logic of capitalism, whereas concrete historical study has to come to terms with the realm of freedom in which necessity is hidden and mystified.

In Chapter 3, it was suggested that Harvey's struggle in the early 1970s to avoid reductionism and determinism, and yet insist upon viewing the economic aspect of society as the substructure of a superstructure which ultimately reflected it, was also a struggle with the tension between freedom and necessity in Marx's writings which Marxists inherit. In the late 1970s, Harvey continued, in his writings on urbanisation under advanced capitalism, the struggle to avoid a mere reductionism. It was in the economic substructure that the key to explanation lay. Harvey believed, for example, that in so far as the three

circuits of capital portrayed 'capital movements into the built environment (for both production and consumption) and the laying out of
social expenditures for the reproduction of labour power', they provided 'the structural links we need to understand the urban process
under capitalism' (Harvey, 1978b, p. 114). He also admitted that his
concentration upon accumulation 'may sound rather "economistic" as
a framework for analysis' (Harvey, 1978b, p. 102). Elsewhere, he noted
the existence of a 'very "economistic" and "reductionist" school within
Marxism' (Harvey, 1978c, p. 35) but, by implication, disassociated himself from it.

In his defence, Harvey argued that his treatment of class struggle as
the 'other side of the coin' from capitalist accumulation (Harvey, 1978b,
p. 101) meant that the analysis was not simply economistic. The 'economic base' in Marxian theory was not the same as the 'economic factor'
in bourgeois social science (Harvey, 1978c, pp. 35-6). However, Harvey
struggled to integrate the dichotomous Marxian class model with the
range and variety of classes recognisable in the empirical situation he
studied. For example, he developed the notions of subjective classes
(Harvey, 1974b, p. 250), housing classes (Harvey, 1974b, pp. 252-4,
1976a, pp. 272-3) and class factions (Harvey, 1976a, p. 265) but argued
for the existence of the objective class structure of capital and labour
(Harvey, 1974b, p. 250) and asserted that conflicts in the 'living place'
of labour were 'mere reflections' of the underlying tension between
capital and labour (Harvey, 1976a, pp. 289, 294). This struggle to integrate Marx's model with the reality of the empirical situation is one of
the dimensions of the struggle Harvey faced in attempting to integrate
theory and concrete historical reality in his analyses.

The second aspect of Marx's method presented in *Social Justice and
the City* was constructivism, the view that the epistemological subject
both structures and is structured by the object. It has been suggested
earlier that, in his analysis of urbanisation under advanced capitalism,
Harvey has approached the object of his research with the categories
of Marxian theory already in mind. It has also been suggested that his
research has not sought to test the theory but rather to confirm or
illustrate it. It could therefore be concluded that Harvey has not
allowed his analyses to be structured by the epistemological object
to the extent prescribed by the constructivist position. However, his
writings tend to report a particular analysis rather than detail the process by which the analysis was finalised. It could be argued that Harvey
has tested his theoretical propositions on urbanisation, found some of
them inadequate and only reported the successful ones. Many academic

publications and reports appear to follow this practice. The process itself of formulating and testing propositions is often not reported and the resulting analyses appear out of the context of their origin, as if they had occurred spontaneously to the researcher in their final form. However, it is difficult to envisage the way in which some of the more basic theoretical propositions of Harvey's Marxist analysis may, in fact, be tested. For example, the proposition that capital accumulation is the basic underlying process of capitalism is very difficult to falsify. Every empirical situation may be explained in relation to this process because, in the first place, it is assumed that everything in capitalism is relationally linked to everything else and, secondly, the introduction of other variables may be used to account for seeming contradictions. Those actions of the State which seem to be in the interests of the working class, for instance, prove upon further analysis to be primarily in the interest of continued capital accumulation (Harvey, 1976b, pp. 81-2). All theoretical positions share this problem in that certain assumptions must be made to shape and guide research. However, all theoretical positions should not be held dogmatically but be open to debate and to empirical evidence.

In his writings in the late 1970s, Harvey has not been unaware of the tension between freedom and necessity as it has been manifested in the issues concerning reductionism, the relationship between theory and practice and between theoretical and concrete historical analyses. He has consciously been struggling to resolve them. The analysis provided in this chapter suggests that his work on urbanisation under capitalism has tended towards the necessity pole in each of the various dichotomies, that is, it tends towards economic reductionism, theoretical idealism and the imposition of predetermined categories. However, as both Gouldner (1980, pp. 34, 54) and Dooyeweerd (1953, p. 210, 1960, pp. 45-51) have pointed out, each pole of the dichotomy presupposes the other as well as excludes it. There are therefore elements of non-reductionism and concrete historical analyses in Harvey's writings, for example in 'Monument and Myth' (Harvey, 1979). There is also an indication of his involvement in a 'practical' issue, the Baltimore Rent Control Campaign, however much its 'revolutionary' character might be questioned. The challenge for Harvey in the 1980s is to show the extent to which a resolution of the various dichotomies is possible within the Marxist tradition.

In commenting upon the hegemony of the 'bourgeois framework for organizing knowledge' (Harvey, 1978c, p. 35), Harvey noted that my employment prospects are almost entirely enclosed within the professional framework of geography; yet my colleagues in this field typically dismiss my work as "not geography", but political economy, sociology, etc. But I do not possess the professional credentials to be considered a bona fide critic in these fields (Harvey, 1978c, p. 44, note 1).

The fact that Harvey did not conform to the 'professional framework of geography' in the 1970s has had a profound influence upon the discipline. Some of his most significant contributions have been to bring the capitalist character of urbanisation in the advanced capitalist countries into focus, to introduce important questions relating to the ideological and political bias of so-called 'scientific' urban geographic studies and to assist in the consideration of a greater range of philosophical and methodological perspectives than had previously been admitted.

## Notes

1. Harvey has been an editorial adviser for the *International Journal of Urban and Regional Research* since its founding in 1977 and became a member of the editorial board of *Antipode* in 1979.

2. In his writings in the late 1970s, Harvey frequently referred to 'capital' and 'labour', as well as 'finance capital' and various 'factions of capital', when dealing with particular social or economic classes. This depersonalised, almost reified, mode of language will be reproduced in the following examination of Harvey's writings in order to reflect his terminology. It should not be assumed that the author agrees with such usage.

3. Harvey used the term 'Marxian' to describe those theories which Marx himself formulated and applied the term 'Marxist' to theories developed by those scholars seeking to extend or modify Marx's formulations.

4. The use of the impersonal term 'which', instead of 'who', is a deliberate attempt to mirror Harvey's terminology.

# 5 CONCLUSION

In 1972, Harvey was presented the Gill Memorial Award by the British Royal Geographical Society 'for contributions to theoretical geography' (*Geographical Journal*, 138, 1972, p. 271). Then, in 1980, the Association of American Geographers conferred on Harvey an Honours Award, citing his

> important scholarly contributions in developing the philosophical bases of analytical and behavioral studies in human geography and in providing alternative explanations of urban geography phenomena based on the tenets of classical political economy (*Association of American Geographers, Newsletter*, 15(3), 1980, p. 3).

The presentation of these awards indicates professional recognition of Harvey's contributions to the development of two quite different philosophical and methodological approaches in Anglo-American geography, namely, geography as spatial science and Marxist geography. It also confirms Harvey's influential role in the development of the discipline over the last 20 years.

In this chapter, conclusions concerning the phases of Harvey's geography and the various philosophical stances he adopted will be summarised. Continuities and discontinuities in his writings will be reviewed. An assessment will be made of the implications of our analysis of Harvey's geography for the study of change in geography. Finally, the relationships between philosophy, methodology and geographical research as apparent in Harvey's work will be examined.


## David Harvey's Geography, 1961-1981

In the Introduction, a number of questions were posed regarding the philosophical positions evident in Harvey's publications over the last two decades. In the course of our analysis, it was shown that Harvey's doctoral thesis fell between the older traditional geography and an emerging spatial science geography. Harvey was unhappy with the restrictions of the former but not yet ready to enter fully into the spirit of the latter. In contributing towards the development of spatial science geography over the 1960s, Harvey eventually explored the logical empiricist philosophy of science on the basis of which he advocated a model-based paradigm, utilising an instrumentalist conception of laws and theories (Harvey, 1969a). After this preoccupation with the *form* of scientific explanation in geography, Harvey turned his attention to the *content* of geographical theory and, in 1972, he rejected logical empiricism in favour of a Marxist position (Harvey, 1972c). In *Social Justice and the City*, the Marxist philosophical position was presented by Harvey in terms of Marx's method which utilised an operational structuralist ontology and a constructivist epistemology (Harvey, 1973a). In this manner, Harvey attempted to convey the Marxist perspective as a materialist, but not determinist or reductionist, approach. While this approach was largely unmodified throughout his writings in the late 1970s, Harvey found it necessary to draw upon more concrete ontological elements of Marx's views, especially the centrality to capitalist society of capital accumulation and class struggle.

Throughout the two decades of Harvey's writings, and despite his radical philosophical and methodological reorientation, there are at least four major continuities that emerge. The most fundamental is a belief in the power of rational analysis. Other continuing concerns are the relationship between form and process, the importance of general theory and the transcendence of disciplinary boundaries in search of satisfying explanations.

In considering Harvey's faith in the power of rational analysis, we may begin by noting that he himself pointed to the materialist base and analytic method common to both logical empiricism and Marxism (Harvey, 1972c, pp. 10-11). A recurring reference in Harvey's work in the late 1960s is to the *power* of scientific analysis and of models and theories. In *Explanation in Geography*, he referred to the 'fantastic power of the scientific method' (Harvey, 1969a, p. vi). But in the face of the demonstrable inability of spatial science geography to deal with contemporary social problems, Harvey turned to what he had come to consider to be the more powerful, critical and realistic approach of Marxism. The alternatives of phenomenology and logico-linguistics were seen to be 'softheaded' whereas logical empiricism had in turn been seen as 'erroneous hardheadedness' (Harvey, 1972f, pp. 326-7). In 1978, Harvey wrote that he wished to be regarded 'primarily as a *scientist* seeking a comprehensive understanding of the world in which we live' (Harvey, 1978c, p. 28). To Harvey, Marxism retained the hardheaded scientific approach of logical empiricism but went further than logical empiricism in embracing and encouraging revolutionary and humanising change. Furthermore, it is of significance to note that both *Explanation in Geography* and *Social Justice and the City* have sketches by Escher embroidering their dust-jackets and, in the words of Ley, 'the rationally ordered elegance of Escher's world is shared both by positivism and Marxism' (Ley, 1978, p. 48). Such symbolism reflects Harvey's assumption that rational analysis is capable of truly describing and explaining reality, despite reality's manifest complexity.

Harvey also believed that the results of rational analysis could guide the search for either the amelioration (Harvey, 1970a, 1971, 1972b) or the revolutionary transformation (Harvey, 1972c, 1973a, Chapters 5 and 6, 1973b, 1974c, 1974d) of pressing social and economic problems. His rejection of logical empiricism and liberalism may be seen as reflecting not a loss of faith in rational analysis but more as a rejection of their lack of both an ethical commitment to, and a clear-cut programme for, dealing with the injustices and inequalities of capitalism. To Harvey, Marxism offered a powerful and realistic analysis of capitalist society, a committed perspective and a radical programme for humanising change.

Tensions and contradictions have been identified in both the logical empiricist and Marxist writings of Harvey. Is there any relationship between these two sets of problems? It could be argued that common to the logical empiricist and Marxist modes of analysis is the belief that certainty resides in a scientific objectivity supposedly guaranteed by the application of a particular method. The two differ, of course, over the character of the proper method, and accuse each other of subjectivity and bias. However, both overlook the subjective character of their own approach and devalue non-scientific human activity and experience. In his attempt to exclude subjectivity from the scientific method of logical empiricism, Harvey postulated the separation of methodology from philosophy. But the separation could not be maintained. In attempting to exclude subjectivity from Marx's method, the only method capable of uncovering (demystifying) what is really happening in society, Harvey was forced to differentiate a realm of necessity, the theoretical, from a realm of freedom, the concrete historical. But the differentiation became another separation and the reintegration of the two became a major challenge yet to be adequately met by Harvey. Both logical empiricism and Marxism project upon reality an abstract, rationalist character having its origin more in their own positivist view of science than in the integral and multi-faceted nature of reality itself. Consequently, neither mode of analysis is able to deal adequately with reality, and tensions and contradictions develop within their research programmes.

The resolution of the tensions and contradictions within logical empiricism and Marxism partly lies in a recognition of the limited and abstract character of scientific knowledge, which has its roots in an unavoidable human subjectivity. The desire to locate certainty within the bounds of science needs to be seriously questioned and the significance of non-scientific forms of experience and knowledge rediscovered (Dooyeweerd, 1960, pp. 1-38, 45-51; Schuurman, 1977).[1]

The second major theme which recurs throughout Harvey's work is a concern with the relationship between form and process. Harvey received from traditional regional geography a view of the distinction between form and process which had led to the definition of geography as the study of form alone. In his doctoral research, Harvey rejected this restriction and went on to examine the processes underlying the patterns of agricultural change in Kent during the nineteenth century. Later in the 1960s, Harvey examined statistical techniques analysing spatial patterns from which inferences regarding processes had been attempted. He warned of the dangers of this type of inference and of inferring form from process as well. In an article in 1967, Harvey identified the fundamental concern of human geography as being the analysis of 'spatial processes' and indicated that geography had to derive theories about these processes from other disciplines (Harvey, 1967c). In *Explanation in Geography*, he defined general theory in geography as exploring the links between indigenous theories of spatial form and derivative theories of temporal process (Harvey, 1969a). Thus, Harvey redefined the traditional view of the relationship between form and process to place both at the centre of geography's subject-matter. Later, in *Social Justice and the City*, Harvey presented the relationship between spatial form and social process as one of interpenetration, rather than, as previously, one of space containing processes. Social processes *are* spatial (Harvey, 1973a). By the late 1970s, in his analysis of urbanisation under advanced capitalism, Harvey had largely turned his attention to the examination of economic, social and political processes, although he retained a keen awareness of the spatial dimension of such processes.

The third major theme of continuity in Harvey's work is that of the importance of general theory. Although he initially rejected the construction of theory in the context of his doctoral research, Harvey went on, in the late 1960s, to adopt the view that the development of theory was central to a scientific geography. As early as 1966, he had emphasised the importance of a general theory of agricultural location in linking economic and behavioural models (Harvey, 1966a). In *Explanation in Geography*, he presented theory as a hypothetico-deductive system made up of a hierarchy of theoretical statements at the top of which was a general theory. From this general theory, all the other theoretical statements were to be deduced. When, in the 1970s, Harvey rejected this logical empiricist view, he did so in favour of what could be called a general theory, namely, Marxism. Marxism not only provided, within an overarching unity, the satisfying theoretical content that Harvey was seeking, but also an alternative method to that of logical empiricism. Furthermore, Marx's theory acted as a guide to Harvey's analyses and was used as a hypothetico-deductive system, to generate hypotheses which could be explored in the context of urbanisation under advanced capitalism. Thus, throughout most of the 1960s, Harvey was advocating the development of general theory in geography and, in the 1970s, he discovered that Marxism provided him with just such a theory.

The fourth major theme recurring throughout Harvey's writings relates to the call for and adoption of an approach which went beyond the boundaries of any one discipline. For Harvey in the 1960s, a geography that did not draw upon any other discipline could not offer any explanation for the spatial patterns it analysed. Human geography had to have recourse to theories developed in other social sciences such as economics, sociology and psychology. Harvey himself encouraged the development of behavioural geography, seeing it as adding valuable insight into the decision-making processes behind the more economic locational analyses. As we noted above, in *Explanation in Geography*, Harvey defined geography's general theory as exploring the relationship between indigenous theories of spatial form and derivative theories of temporal processes. Thus, even within the definition of geography implicit in his indication of the nature of general theory in geography, Harvey encompassed theories derived from other disciplines.

After the publication of *Explanation in Geography*, Harvey conducted interdisciplinary research on the city. In one article, for instance, he explored the relationship between geography and sociology (Harvey, 1970a). But after the adoption of a Marxist mode of analysis, Harvey viewed disciplinary boundaries as counter-revolutionary, rendering research impotent to bring about revolutionary change in the face of the capitalist State. For him, interdisciplinary research had to be replaced by meta-disciplinary research because the former assumed the prior existence of separate disciplines. Meta-disciplinary research approached reality directly and not through a disciplinary framework (Harvey, 1972e).

Later, in *Social Justice and the City* and in articles on Marx's method,

Harvey asserted that operational structuralism was the only kind of method capable of dealing with the complexities of reality in an integrative fashion. This implied that it was the only method that could be successfully utilised in meta-disciplinary research. Consequently, in his Marxist analyses, Harvey concentrated upon processes usually seen to fall within the domain of economics, political science and sociology, although there were significant points of contact with traditionally geographical issues. In the application of operational structuralism to an analysis of the Baltimore housing market, for example, Harvey identified both a geographical (spatial) and a social structure, in which class-monopoly rent was realised (Harvey and Chatterjee, 1974; Harvey, 1974b). In another article, he examined Marx's location theory in relation to the theory of capital accumulation (Harvey, 1975d). Harvey's geographical concerns did not enable him to find a home within the literature of any other single discipline, however.

There are perhaps two major discontinuities in Harvey's work, the second more significant than the first. The first discontinuity is that between the subject-matter of Harvey's writings in the early 1960s, the late 1960s and the 1970s. In the early 1960s, he was concerned with the study of agriculture in Kent in the nineteenth century (Harvey, 1961, 1963), going on in 1966 to review the models and theories that had been utilised in the analysis of agricultural land-use patterns (Harvey, 1966a). A number of methodological articles followed in the late 1960s, culminating in the logical empiricist analysis of scientific methodology in *Explanation in Geography*. Then, in the 1970s, perhaps occasioned partly by his move to Baltimore and partly by a desire to conduct research in what had become one of the most significant fields of spatial science geography, Harvey turned his attention to the study of cities, ghettos and urban housing markets. As with the earlier research on agriculture and with some of his methodological studies in the 1960s, Harvey's research in the 1970s dealt with social and economic subject-matter. At various times, Harvey was also concerned with historical analysis (Harvey, 1961, 1973a, pp. 246-61, 1979). Both of these interests tend to make the transition to Marxism much more likely than if Harvey had followed up the more ahistorical psychological and symbolic aspects of behavioural geography that he had surveyed in an article published in 1969 (Harvey, 1969b). Exceptions to Harvey's urban studies in the 1970s were a number of papers exploring Marx's methodology in relation to the population-resources debate (Harvey, 1973b, 1974c, 1974d).

The second and more significant major discontinuity in Harvey's writings is that between the two philosophical positions adopted by him in the 1960s and the 1970s, namely logical empiricism and Marxism. The last three continuities in Harvey's writings examined, namely, concern with the relationship between form and process, general theory and the transcendence of disciplinary boundaries, are all given quite different meanings by Harvey according to the philosophical assumptions held at different times. The philosophical discontinuity in Harvey's thought is therefore reflected in discontinuities in the meaning assigned even to concerns present throughout the range of his work. One further discontinuity reflecting the basic philosophical one is that regarding Harvey's views of the relationship between philosophy, methodology and geographical research, views which will be examined in the concluding section of this chapter.

### David Harvey and Philosophical and Methodological Change

This study has sought to present an internal history of David Harvey's thought as it is expressed in his published writings, and to document the changes in his philosophical and methodological views. Inasmuch as change in geography has been either advocated or analysed from both Kuhn's and Popper's viewpoints, it is important to assess their views in the light of the examination of Harvey's publications. Before briefly considering their views, however, it must be noted that Kuhn and Popper were referring primarily to the physical sciences and not to the social and behavioural sciences, nor to the work of only one researcher. Any generalisations about geography as a discipline based upon the conclusions of this study of Harvey will need to draw upon other more comprehensive analyses as well. The nature of change in a discipline cannot be ascertained solely from the analysis of change in the writings of one geographer, although such an analysis can raise important questions about and provide essential, if limited, insight into the nature of change.

Kuhn distinguished between normal and revolutionary science. Normal science is conducted within the framework of a ruling disciplinary matrix, solving problems which have meaning in relation to the disciplinary matrix. Normal science is thus cumulative and largely unaffected by major theoretical debates. Revolutionary science, on the other hand, accompanies the overthrow of one disciplinary matrix by another, is innovative, progress is discontinuous and a new way of viewing the world results (Kuhn, 1962, 1970c). For Popper, however, science is continually revolutionary and innovative, its progress achieved through a series of conjectures and refutations. Science is never dogmatic nor uncritical as implied by Kuhn's description of normal science (Popper, 1963, 1970, 1972a).

The internal history of Harvey's thought, as presented in Chapters 2, 3 and 4, tends to reflect the Kuhnian rather than the Popperian picture of scientific change. Harvey's doctoral thesis has been represented as falling between regional and spatial science geography. His work from about 1965 to 1969 could be characterised as normal science within the spatial science disciplinary matrix. Then followed a period of rapid and radical change. Three articles concerned with issues of relevance from a liberal viewpoint were presented by Harvey at various conferences in 1969, 1970 and 1971, each being published a year after its presentation (Harvey, 1970a, 1971, 1972b). It was in 1972 that Harvey published the article in which, to use Peet's (1977, p. 249) term, he made the 'breakthrough' to a Marxist perspective (Harvey, 1972d). This article was soon followed up by three publications, the first, a reply to Gale's review of *Explanation in Geography*, in which Harvey rejected logical empiricism (Harvey, 1972f), the second, a resource paper offering a tentative Marxist approach to the analysis of the city (Harvey, 1972g), and the third, a book review in which Harvey offered an alternative Marxist hypothesis on the question of the origins of urbanism (Harvey, 1972h). From this point on, Harvey conducted what might be termed normal science within the Marxist disciplinary matrix. Harvey's work may thus be presented in Kuhn's terms as two periods of quite distinct types of normal science separated by a period of rapid and radical change. Scientific progress appears to be discontinuous between the two periods of largely cumulative normal science.

In relation to Popper's view that scientific progress consists of a sequence of conjectures and refutations, there being some measure of continuity between theories, it is relevant to consider Harvey's rather ambiguous views on the character of revolutionary theory. On the one hand, he believed that any theory, even a positivist one, could be revolutionary if it entered revolutionary practice (Harvey, 1972e, p. 41) whereas, on the other hand, he advocated the development of alternative Marxist theories which incorporated and 'enriched' non-Marxist theories 'through the integrating power of higher order concepts' (Harvey, 1973b, p. 39). In his Marxist analysis of urbanisation under advanced capitalism in the late 1970s, Harvey followed the second alternative, which has in common with Popper the notion that there is at least some continuity between two theories where one replaces or supersedes the other. However, Popper's views do not take into account the differences between theories developed within different philosophical traditions. Harvey noted at one point, for instance, that location theory has 'no meaning independently of our concepts of rent, capital and labor' (Harvey, 1973c, p. 87), comparing the location theories derived from neo-classical marginalist economics with those derived from Marxism. Thus, contrary to Popper, there may be radical discontinuity between two location theories, one of which appears to subsume the other and both theories using similar terms, if they arise from different philosophical traditions.

Johnston (1979) has suggested that no single disciplinary matrix has dominated Anglo-American human geography since 1945.

> Rather there have been several competing for a stable position, if not dominance, both within the discipline and beyond. Each matrix has its own branches with their particular exemplars and their leaders who chart progress and seek influence over the whole discipline. At times, the number of branches ... may suggest anarchy ... There is no consensus, no paradigm-dominance, only a series of mutual accommodations which reflects the liberal democratic societal setting of modern Anglo-American human geography (Johnston, 1979, p. 188).

Johnston emphasised the important role of 'iconoclasts', geographers who, usually with a secure position in the academic career system, destroy beliefs, methods and assumptions that have become established in the discipline, create new exemplars and gain followers. A new branch of the discipline develops, which may come to dominate the disciplinary matrix or eventually replace it (Johnston, 1979, pp. 20, 185). Harvey played an iconoclastic role in the 1970s in his criticism of the logical empiricist assumptions of spatial science geography, in his 'breakthrough' to Marxist geography and through the influence of *Social Justice and the City* upon other geographers. However, to pursue this line of analysis is to go beyond both the confines of an internal history of Harvey's thought and the material presented in this book. A consideration of Harvey's role in the recent history of Anglo-American human geography and a more detailed examination of the relevance of Kuhn's and Popper's views to progress in geography must have recourse to material on the external social, economic and political conditions prevailing in Anglo-America. Furthermore, the reasons for Harvey's turn to Marxism in the early 1970s cannot be discovered simply through an analysis of his writings. More needs to be known about the man and his motives. Why did Harvey move to the United States? What experiences did he have there that may have influenced his decision to consider Marx's works? Questions such as these need to be pursued if we are to understand more fully the reorientation in Harvey's thought. An internal history of his writings provides only the foundation for further research.

## Philosophy, Method and Geographical Research

In Chapter 1 it was indicated that in this concluding chapter we would consider the relationship between philosophy, methodology and geographical research as apparent in Harvey's writings and, in particular, assess Harrison and Livingstone's (1980) views on the role of philosophical presuppositions in research. The majority of Harvey's articles published in the 1960s were methodologically oriented. In *Explanation in Geography*, Harvey made the point of separating methodology from philosophy. Methodology examined scientific explanation as a formal procedure, demonstrating the form of analysis which ought to be followed if explanation was to be rigorous and logically sound. Philosophy examined the objectives of geographical research which, according to Harvey, could only be disputed on grounds of belief. The objectives a geographer chose to pursue were dependent upon his own individual values. The adoption of a methodological procedure did not entail the adoption of a corresponding philosophical position, although a philosophical position implied a methodology. Thus, in the late 1960s, Harvey believed that methodology was objective and communal, in that it examined standards that ought not to be ignored by any geographer, whereas philosophy was an individual matter and dependent upon subjective foundations.

It is noteworthy that the view of the relationship between philosophy and methodology put forward by Harvey in *Explanation in Geography* derived from the logical empiricist *philosophy* of science and that Harvey himself failed to maintain the separation of philosophy and methodology in the book when, for example, he advocated that a more concise philosophy of geography would be promoted by the adoption of logical empiricist methodology. Furthermore, in the preface to *Explanation in Geography*, Harvey admitted that the methodological changes in geography over the 1960s had meant that the philosophy of geography had to change as well. Harvey soon realised the impossibility of separating philosophical and methodological considerations (Harvey,

1972f) and, in *Social Justice and the City*, he presented what he referred to as the ontological and epistemological aspects of Marx's method. Harvey came to view methodology as being based upon philosophical considerations.

In publications in 1973 and 1974, Harvey compared Marx's method with that of Malthus and Ricardo, seeking to demonstrate that, within the context of the population-resources debate, each method generated its own particular set of conclusions. Harvey consequently attacked the pretended ideological neutrality and objectivity of the so-called scientific method. In the late 1970s, he went on to apply Marx's method to the analysis of urbanisation under advanced capitalism. Harvey thus rejected his earlier views on the relationship between philosophy, methodology and research. He came to recognise that a method, on the one hand, necessarily entailed ontological and epistemological presuppositions and, on the other, generated the specific conclusions of research. This position is in general accord with the presuppositional hierarchy postulated by Harrison and Livingstone in their brief and largely prescriptive article (Harrison and Livingstone, 1980, pp. 26-7). As illustrated in Figure 1.1 (p. 9), they indicated the influence of ontological presuppositions about the nature of reality and the sources of knowledge upon epistemological presuppositions which delimit the domain of enquiry and specify legitimate questions for research. However, part of what is identified as ontological by Harrison and Livingstone, namely, the sources of knowledge, is referred to by Harvey in Chapter 7 of *Social Justice and the City* as epistemological (Harvey, 1973a, pp. 297-8). For Harrison and Livingstone, epistemological presuppositions determine, or structure, the methodology to be utilised in research, that is, the organisation of the analysis and the type of analytical techniques and instruments to be used. Again, this largely reflects Harvey's views on methodology as presented in Chapter 7 of *Social Justice and the City*.

There are, however, two major differences between Harrison and Livingstone's and Harvey's views of the role of philosophical presuppositions in research. First, Harrison and Livingstone believed that ontological presuppositions were based upon cosmological presuppositions about the origin of reality. Harvey does not discuss cosmological presuppositions in any of his publications, going no further than an examination of ontological issues. But it could be argued that cosmological presuppositions regarding the origin of reality may be detected in Marx's writings. In the Introduction to his *Contribution to the Critique of Hegel's Philosophy of Law*, Marx developed a radical humanistic starting point.

Criticism of religion is the premise of all criticism ... The basis of irreligious criticism is: *Man makes religion*, religion does not make man ... The criticism of religion disillusions man to make him think and act and shape his reality like a man who has been disillusioned and has come to reason, so that he will revolve round himself and therefore round his true sun. Religion is only the illusory sun round which man revolves as long as he does not revolve round himself (Marx and Engels, 1975, pp. 175-6).[2]

In the *Economic and Philosophic Manuscripts of 1844*, Marx addressed the question of the origin of mankind and of nature in the light of his critique of religion.

A *being* only considers himself independent when he stands on his own feet; and he only stands on his own feet when he owes his *existence* to himself. A man who lives by the grace of another regards himself as a dependent being. But I live completely by the grace of another if I owe him not only the maintenance of my life, but if he has, moreover, *created* my *life* – if he is the *source* of my life. When it is not of my own creation, my life has necessarily a source of this kind outside of it ... But since for the socialist man the *entire so-called history of the world* is nothing but the creation of man through human labour, nothing but the emergence of nature for man, so he has the visible, irrefutable proof of his *birth* through himself, of his *genesis* (Marx and Engels, 1975, pp. 304-5).

Thus, for Marx, man created himself through human labour and in these terms may be said to have his origins in neither the creative act of a transcendent god nor the mechanical processes of nature (Van der Hoeven, 1976, pp. 83-6, 108). Nature, on the other hand, may be said to be the creation of man in that it finds its meaning and significance in relation to man's self-creation. It could be argued that such cosmological views on the origin of humankind and nature are presuppositional to Marx's later writings. Indeed, such concepts in Marx's political economic analyses as the forces of production and the labour theory of value presuppose a view of the relationship between people and nature which is based upon Marx's cosmological presuppositions. Ollman (1971, pp. 73-127) and Van der Hoeven (1976), among others, have examined the influence of cosmological presuppositions upon Marx's views. If such scholars are correct, and if Harrison and Livingstone's presuppositional hierarchy is valid, then the ontological views that Harvey adopted from Marx presuppose such cosmological assumptions.

The second difference between Harrison and Livingstone's and Harvey's views on the role of presuppositions in research relates to what the former call 'disciplinary' presuppositions. These define the aspect of reality to be investigated by the researcher and, as Figure 1.1 (p. 9) has shown, these were seen by Harrison and Livingstone to be structured by ontological presuppositions and to contribute to the definition of epistemological presuppositions. In the late 1970s, Harvey advocated a meta-disciplinary approach to research. The city, for example, should be studied without the interference of disciplinary frameworks. But, in such a case, disciplinary presuppositions are simply replaced by presuppositions designating the subject-matter of research. Harvey, for instance, was not concerned with a microscopic analysis of the city but implicitly approached it at a certain scale and with definite interests in mind.

If the distinctions made by Harrison and Livingstone concerning ontology, epistemology and methodology are accepted as valid, Harvey's work at a general level may be characterised in the following manner. In his doctoral research, Harvey was largely concerned with historical economic research in the Hartshornian tradition, the cosmological, ontological and epistemological presuppositions of which were of no concern to him. However, he believed that the disciplinary presuppositions of geography as stated by Hartshorne were too restrictive and Harvey abandoned them in order to study the processes behind locational change. The methodological presuppositions entailed in the study were those of traditional regional geography extended by several simple regression analyses. Subsequently, Harvey became involved in an examination of the methodological presuppositions of spatial science geography which led him to consider epistemological and disciplinary presuppositions, particularly those consistent with a logical empiricist perspective. In the early 1970s, Harvey went on to consider the ontological presuppositions of spatial science geography, especially those relating to the character of society. Finding the Marxist perspective more realistic and attractive than that of liberalism, Harvey adopted Marxist ontological presuppositions which immediately led him to an examination of the disciplinary, epistemological and methodological presuppositions of Marxism which are encapsulated in his outline of Marx's method. Harvey then analysed urbanisation under advanced capitalism on the basis of these Marxist presuppositions, arriving at a framework which may be described, in the words of Harrison and

Livingstone, as 'both directed and structured by these presuppositional influences' (Harrison and Livingstone, 1980, p. 27).

However, Harvey was not necessarily aware of all the assumptions contained in his work. He has not demonstrated, for example, an awareness of cosmological presuppositions and his experience during the early 1970s revolved around a growing awareness of the assumptions inherent in logical empiricism. Of course, such assumptions were implicit in his writings whether Harvey was aware of them or not. In general, this study tends to affirm the existence of tacit philosophical presuppositions in Harvey's research. It also indicates that philosophical presuppositions do not act as axioms in a logical argument, as in a hypothetico-deductive system (Harvey, 1969a, p. 36). Rather, they either prescriptively assign meaning to the concepts used in analysis, not unlike the role of conceptual frameworks (Harvey, 1973a, p. 245), or retrospectively describe the meaning implicit in research which has already been conducted. Harvey's application of Marx's method to the study of urbanisation partly entails the prescriptive assignment of meaning, whereas his critique of logical empiricism in the early 1970s retrospectively described the meaning implicit in his earlier research.

Human geographers have responded to and analysed the recent development of philosophical and methodological pluralism in a variety of ways. Gregory (1978), for example, has attempted to synthesise three types of explanation in order to develop a critical human geography. However, such an approach does not necessarily succeed in producing a superior perspective but may only produce yet another perspective with its own strengths and weaknesses. One of the main weaknesses of such a synthetic human geography is likely to be the tensions between what were originally radically incompatible approaches (Hindess, 1979, p. 351). Johnston (1979, p. 189), on the other hand, believes that human geography is 'branching towards anarchy'. But perhaps what might be called a 'non-relativistic philosophical pluralism' is more appropriate to Anglo-American human geography in the 1980s than an either synthetic or anarchic approach. The integrity of each philosophical perspective ought to be recognised and allowed to develop freely within the discipline. This need not imply a philosophical relativism, that is, that each approach is as 'truthful' or valid as any other. Rather, a non-relativistic pluralism entails critical commitment to that perspective which the researcher finds most in accord with his or her view of reality. It also encourages continual reflective communication between those holding different perspectives. Constructive communication between the various philosophical perspectives in human geography may be developed in a number of ways, one of which is represented by the kind of study undertaken in this book. Harvey's experience is of particular significance in that he played a leading role in the development of two radically different philosophical approaches in human geography, rejecting one perspective in favour of another.

Another way in which communication between philosophical perspectives in human geography may be encouraged is in the development of comparative studies. Harvey's 'The Political Economy of Urbanization in Advanced Capitalist Countries: The Case of the United States' (Harvey, 1975c) was published in a book which contained two other analyses of urban housing markets in the United States, namely, those by Berry and by Nourse and Phares. Berry's essay, after outlining Chicago's dual housing market (based on race), examined the influence of short-term housing swings upon the access of blacks to housing (Berry, 1975). Nourse and Phares analysed the influence of the income variable upon temporal changes in housing values in 15 neighbourhoods within St. Louis County (Nourse and Phares, 1975). The editors commented that each of the three papers represented a new approach to topics which had previously received much attention. They then posed the question of the influence of different analytical frameworks or models upon research. 'Do the models emphasize or obscure the differences in housing market conditions in these three different cities?' (Gappert and Rose, 1975, p. 117).

We may extend this question to ask in what way a Marxist urban geography may be more 'successful' than more conventional positivist approaches in analysing and explaining urbanisation under advanced capitalism. The examination, in Chapters 3 and 4, of Harvey's urban analyses may provide a contribution towards such a comparative study. However, it should be recognised that comparative studies need to probe to the philosophical roots of research. For, as Harrison and Livingstone's presuppositional hierarchy illustrates, research is directed and structured by philosophical influences as well as by its subject-matter and external social, economic and political conditions.

This last point, that research is influenced by philosophical considerations, subject-matter and external conditions, leads us to consider what appears to be one of the most important issues in contemporary Anglo-American human geography. Throughout his work, Harvey was vitally concerned in various ways with the relationship between theory, or consciousness, and material reality. In the 1960s, he encouraged the development of behavioural geography because it provided insight into the decision-making processes overlooked by the economic approach.

It was in considering the application of theory to urban planning that he recognised the *status quo* and counter-revolutionary character of logical empiricist theories and the need to develop a change-embracing theoretical framework. In Chapter 3, it was noted that Harvey viewed revolutionary theory as necessary to revolutionary material change. An important question arising in connection with his analysis of urbanisation under advanced capitalism is the extent to which it is verified in practice by creating revolutionary truth, the view of verification that Harvey adopted in *Social Justice and the City* (Harvey, 1973a, pp. 12, 151). To examine such a question is beyond the scope of this study. But the issue of the relationship between theory and material reality will be central to the problematic of human geography in the 1980s.

The issue of theory's relationship to material reality has variously been stated as the question of the relationship between science and society (Gregory, 1978, p. 168), between academic research and the external environment (Johnston, 1979, p. 23) and between consciousness and existence (Peet, 1980, pp. 91-2). Associated with it is the problem of objectivity and subjectivity, the need to recognise and account for the value character of facts and the inescapable fact of values. Ultimately, self-knowledge and self-critique must lead the way in working towards a satisfactory resolution of such issues. Harvey has provided a useful model in the development of a penetrating critique of his logical empiricist and liberal formulations. But self-critique is founded upon the achievements of bold, committed and mature scholarship. Human geographers concerned for the philosophical future of the discipline therefore look forward to the time when Harvey presents a mature and rounded Marxist synthesis of the theoretical and the concrete historical. Then a more comprehensive study of Harvey's writings and activities will be able to speak meaningfully to the development of human geography in the years ahead.

## Notes

1. Dooyeweerd (1960, pp. 5-6) also argued that Husserl, the founder of modern phenomenology, was fundamentally uncritical with respect to his method and dogmatically held to the autonomy of scientific thought. Thus Husserl participated in the same basic dialectical tension as Marxism and logical empiricism, despite having a distinctly different emphasis in his critique of knowledge.

2. I have used Bottomore's translation in parts of this passage (see Marx, 1963, p. 44).

# APPENDIX: DAVID HARVEY'S WRITINGS, 1961-1981

Harvey's writings are arranged in chronological order of publication. Within each annual period, books are listed first, followed by booklets, contributions to collections of essays, periodical articles, book reviews, and then comments, replies to reviews and so on. Information on date of writing, presentation at conferences and reprinting is also included where available.

## 1961

'Aspects of Agricultural and Rural Change in Kent, 1800-1900', unpublished PhD dissertation, Department of Geography, University of Cambridge, 259pp. Subtitled 'A Study of the Development of the Hop and Fruit Industries in Kent during the Nineteenth Century and the Effect which these Developments Had upon Other Aspects of the Rural Economy'. Submitted in October 1960 and approved by the Board of Research Studies on 31 January 1961

## 1963

'Locational Change in the Kentish Hop Industry and the Analysis of Land Use Patterns', *Transactions and Papers of the Institute of British Geographers*, 33, December, 123-44
Reprinted in:
R.H.T. Smith, E.J. Taaffe and L.J. King (eds) (1968), *Readings in Economic Geography: the Location of Economic Activity*, Rand McNally, Chicago, pp. 79-93. As originally published except for Figure 8, where the $x$ and $y$ axes were reversed
A.R.H. Baker, J.D. Hamshere and J. Langton (eds) (1970), *Geographical Interpretations of Historical Sources: Readings in Historical Geography*, David and Charles, Newton Abbot, pp. 243-64, with supplementary note, pp. 264-5. Figure 8 as in Smith *et al.* (1968)

## 1965

'Simulation Models' in G. Olsson and O. Warneryd (eds), *Meddelande fran ett Symposium i Teoretisk Samhallsgeografi*, Forskningsrapporter fran Kulturgeografiska Institutionen, Uppsala Universitet, No. 1, pp. 47-8. Abstract of a paper presented at a symposium at Uppsala University, Sweden, in April 1965

## 1966

'Theoretical Concepts and the Analysis of Agricultural Land-use Patterns', *Annals of the Association of American Geographers*, 56(2), June, 361-74. Review article, accepted for publication on 15 November 1965
'Geographical Processes and the Analysis of Point Patterns: Testing Models of Diffusion by Quadrat Sampling', *Transactions of the Institute of British Geographers*, 40, December, 81-95. Revised manuscript received on 28 February 1966. An earlier version of this paper was circulated as a research note while Harvey was a visiting lecturer at the Department of Geography, Pennsylvania State University, in 1965-1966

## 1967

'Models of the Evolution of Spatial Patterns in Human Geography' in R.J.Chorley and P. Haggett (eds), *Models in Geography*, Methuen, London, pp. 549-608

*Behavioural Postulates and the Construction of Theory in Human Geography*, University of Bristol, Department of Geography, Seminar Paper, Series A, No. 6, 29pp. Paper to be read at the Third Anglo-Polish Seminar in September 1967

Reprinted in:

*Geographia Polonica*, 18, 1970, 27-45. Some modifications and additions

'Editorial Introduction: the Problem of Theory Construction in Geography', *Journal of Regional Science*, 7(Supplement), Winter, 211-16

## 1968

'Some Methodological Problems in the Use of the Neyman Type A and the Negative Binomial Probability Distributions for the Analysis of Spatial Point Patterns', *Transactions of the Institute of British Geographers*, 44, May, 85-95. Revised manuscript received on 13 October 1967. Originally presented as a discussion paper to the Statistics Study Group of the Institute of British Geographers in January 1967

'Pattern, Process, and the Scale Problem in Geographical Research', *Transactions of the Institute of British Geographers*, 45, September, 71-8. Manuscript received on 8 July 1967. Originally read to a Regional Science Association (British Section) meeting in March 1967 (Harvey, 1967b, references)

## 1969

*Explanation in Geography*, Edward Arnold, London, 521pp. Preface dated March 1969, at Clifton, Bristol. Translated into Russian and published in 1974 as *Nauchnoye Ob"yasneniye v Geografii. Obshchaya Metodologiya Nauki i Metodologiya Geografii* (Scientific Explanation in Geography. A General Methodology of Science and the Methodology of Geography), abridged translation by V. Ya. Barlas and V.V. Golosov, Progress Publishers, Moscow, 502pp

'Conceptual and Measurement Problems in the Cognitive-Behavioral Approach to Location Theory' in K.R. Cox and R.G. Golledge (eds), *Behavioral Problems in Geography: a Symposium*, Northwestern University Studies in Geography, 17, pp. 35-68. This paper was solicited for a special session entitled 'Behavioral Models in Geography' at the annual meeting of the Association of American Geographers in Washington, DC, in 1968, but was unavailable at that time

Reprinted in:

K.R. Cox and R.G. Golledge (eds) (1981), *Behavioural Problems in Geography Revisited*, Methuen, London, pp. 18-42

'Review of A. Pred (1967), *Behavior and Location: Foundations for a Geographic and Dynamic Location Theory*, Part 1', *Geographical Review*, 59(2), April, 312-14

## 1970

'Social Processes and Spatial Form: an Analysis of the Conceptual Problems in Urban Planning', *Papers of the Regional Science Association*, 25, 47-69. Read to a Regional Science Association meeting in November 1969, in Santa Monica

Reprinted in:

D. Harvey (1973), *Social Justice and the City*, Edward Arnold, London, Chapter 1. Some grammatical editing

'Locational Change in the Kentish Hop Industry and the Analysis of Land Use Patterns: Supplementary Note' in A.R.H. Baker, J.D. Hamshere and

J. Langton (eds), *Geographical Interpretations of Historical Sources: Readings in Historical Geography*, David and Charles, Newton Abbot, pp. 264-5. Dated August 1969

1971

'Social Processes, Spatial Form and the Redistribution of Real Income in an Urban System' in M. Chisholm, A.E. Frey and P. Haggett (eds), *Regional Forecasting: Proceedings of the Twenty-second Symposium of the Colston Research Society*, Butterworths, London, pp. 267-300. Presented at the symposium at the University of Bristol in April 1970
Reprinted in:
M. Stewart (ed.) (1972), *The City: Problems of Planning*, Penguin, Harmondsworth, pp. 296-337. Contains pp. 270-300 of original
D. Harvey (1973), *Social Justice and the City*, Edward Arnold, London, Chapter 2. Some grammatical editing

1972

*Society, the City, and the Space-economy of Urbanism*, Association of American Geographers, Commission on College Geography, Resource Paper No. 18, Washington, DC, 56pp
Reprinted in:
S. Gale and E.G. Moore (eds) (1975), *The Manipulated City*, Maaroufa, Chicago, pp. 132-4 and 271-6. Edited versions of pp. 15-16 and 25-8, respectively, of the original
*The Housing Market and Code Enforcement in Baltimore*, with L. Chatterjee, M.G. Wolman, L. Klugman and J. Newman, The Baltimore Urban Observatory, Baltimore, 211pp. Dated July 1972
'The Role of Theory' in N. Graves (ed.), *New Movements in the Study and Teaching of Geography*, Maurice Temple Smith, London, pp. 29-41
'Social Justice and Spatial Systems' in R. Peet (ed.), *Geographical Perspectives on American Poverty*, Antipode Monographs in Social Geography, No. 1, Worcester, Massachusetts, pp. 87-106. Delivered at the special session on 'Geographical Perspectives on American Poverty and Social Well-being' at the annual meeting of the Association of American Geographers in Boston in April 1971
Reprinted in:
M. Albaum (ed.) (1973), *Geography and Contemporary Issues*, Wiley, New York, pp. 565-84
D. Harvey (1973), *Social Justice and the City*, Edward Arnold, London, Chapter 3. Some grammatical editing
S. Gale and E.G. Moore (eds) (1975), *The Manipulated City*, Maaroufa, Chicago, pp. 106-20
'Revolutionary and Counter-revolutionary Theory in Geography and the Problem of Ghetto Formation' in H.M. Rose (ed.), *Geography of the Ghetto: Perceptions, Problems, and Alternatives*, Perspectives in Geography, vol. 2, Northern Illinois University Press, De Kalb, Illinois, pp. 1-25
Reprinted in
*Antipode*, 4(2), July 1972, 1-13
D. Harvey (1973), *Social Justice and the City*, Edward Arnold, London, Chapter 4. Some grammatical editing, with an additional section being an edited version of the second half of Harvey's reply in *Antipode*, 4(2), to comments on the article (Harvey, 1972e)
'Review of P. Wheatley (1971), *The Pivot of the Four Quarters: a Preliminary Enquiry into the Origins and Character of the Ancient Chinese City*', *Annals of the Association of American Geographers*, 62(3), September, 509-13

'On Obfuscation in Geography: a Comment on Gale's Heterodoxy', *Geographical Analysis*, 4(3), July, 323-30
'A Commentary on the Comments', *Antipode*, 4(2), July, 36-41. Harvey's reply to six comments on 'Revolutionary and Counter-revolutionary Theory in Geography and the Problem of Ghetto Formation'

### 1973

*Social Justice and the City*, Edward Arnold, London, 336pp. Introduction dated January, 1973, at Hampden, Baltimore
*A Question of Method for a Matter of Survival*, University of Reading, Department of Geography, Geographical Paper No. 23, 44pp. Originally delivered as the annual Norma Wilkinson Memorial Lecture at the University of Reading on 18 June 1973. Also presented at the annual meeting of the Association of American Geographers in 1973 at Atlanta
'A Comment on Morrill's Reply', *Antipode*, 5(2), May, 86-8. Follows Morrill's comments on 'Revolutionary and Counter-revolutionary Theory in Geography and the Problem of Ghetto Formation'

### 1974

*FHA Policies and the Baltimore City Housing Market*, with L. Chatterjee and L. Klugman, The Baltimore Urban Observatory, Baltimore, 116pp. Dated April 1974. A preliminary draft, entitled *Effects of FHA Policies on the Housing Market in Baltimore City*, was prepared in August 1973
'What Kind of Geography for What Kind of Public Policy?', *Transactions of the Institute of British Geographers*, 63, November, 18-24. Originally presented at the annual conference of the Institute of British Geographers in Norwich in January 1974. Manuscript received on 6 March 1974
'Absolute Rent and the Structuring of Space by Governmental and Financial Institutions', with L. Chatterjee, *Antipode*, 6(1), April, 22-36
'Class-monopoly Rent, Finance Capital and the Urban Revolution', *Regional Studies*, 8(3/4), November, 239-55. Manuscript received on 6 January 1973; in revised form on 20 March 1974. Also published under same title as University of Toronto, Department of Urban and Regional Planning, Papers on Planning and Design No. 4 (1974)
Reprinted in:
S. Gale and E.G. Moore (eds) (1975), *The Manipulated City*, Maaroufa, Chicago, pp. 145-67
'Population, Resources, and the Ideology of Science', *Economic Geography*, 50(3), July, 256-77. Also published as 'Ideology and Population Theory' in *International Journal of Health Services*, 4(3), 1974, 515-37. Manuscript submitted for publication on 7 March 1974
Reprinted in:
R. Peet (ed.) (1977), *Radical Geography: Alternative Viewpoints on Contemporary Social Issues*, Methuen, London, pp. 213-42
S. Gale and G. Olsson (eds) (1979), *Philosophy in Geography*, Reidel, Dordrecht, pp. 155-86
'Discussion', with B.J.L. Berry, *Antipode*, 6(2), July, 145-9. Follows Berry's review of Harvey's (1973) *Social Justice and the City*

### 1975

'Class Structure in a Capitalist Society and the Theory of Residential Differentiation' in R. Peel, M. Chisholm and P. Haggett (eds), *Processes in Physical and Human Geography: Bristol Essays*, Heinemann, London, pp. 254-72
'The Political Economy of Urbanization in Advanced Capitalist Countries: the

Case of the United States' in G. Gappert and H.M. Rose (eds), *The Social
Economy of Cities*, Sage Publications, Beverley Hills, pp. 119-63. Also published by the Center for Metropolitan Planning and Research, Johns Hopkins
University, in 1974. It was presented at the eighth World Congress of Socio-
logy in Toronto in 1974. It appears to be an extended version of 'Government
Policies, Financial Institutions and Neighbourhood Change in U.S. Cities',
which was published in 1977 but first presented in 1974. In 1976, it was trans-
lated into French and published as 'L'Economie Politique de l'Urbanisation
aux Etats-Unis', *Espaces et Sociétés*, 17, 5-41
'The Geography of Capitalist Accumulation: a Reconstruction of the Marxian
Theory', *Antipode*, 7(2), September, 9-21
Reprinted in:
R. Peet (ed.) (1977), *Radical Geography: Alternative Viewpoints on Contemporary Social Issues*, Methuen, London, pp. 263-92
'Some Remarks on the Political Economy of Urbanism', *Antipode*, 7(1), February,
54-61. Report on a paper presented at a seminar on Marxist geography, organ-
ised by the Union of Socialist Geographers, in November 1974, at Clark Uni-
versity, Worcester, Massachusetts. Draws upon the material also used in 'The
Political Economy of Urbanization in Advanced Capitalist Countries: the Case
of the United States' (1975)
'Review of B.J.L. Berry (1973), *The Human Consequences of Urbanisation*',
*Annals of the Association of American Geographers*, 65(1), March, 99-108
'Review of J. Foster (1974), *Class Struggle and Industrial Revolution – Early
Industrial Capitalism in Three English Towns*', *Journal of Historical Geo-
graphy*, 1(1), January, 109-11

1976

'Labor, Capital, and Class Struggle around the Built Environment in Advanced
Capitalist Societies', *Politics and Society*, 6(3), 265-95
Reprinted in:
K.R. Cox (ed.) (1978), *Urbanization and Conflict in Market Societies*, Methuen,
London, pp. 9-37
A. Giddens and D. Held (eds) (1982), *Classes, Power, and Conflict: Classical
and Contemporary Debates*, University of California Press, Berkeley and
Los Angeles, pp. 545-61. Contains pp. 265-79 and 288-93 of original article
'The Marxian Theory of the State', *Antipode*, 8(2), May, 80-9

1977

'Government Policies, Financial Institutions and Neighbourhood Change in U.S.
Cities' in D.R. Deskins, G. Kish, J.D. Nystuen and G. Olsson (eds), *Geographic
Humanism, Analysis and Social Action: Proceedings of Symposia Celebrating
a Half Century of Geography at Michigan*, Michigan Geographical Publications
No. 17, pp. 291-320. Delivered at the symposium in 1974. Contains much
material in common with 'The Political Economy of Urbanization in Advanced
Capitalist Countries: the Case of the United States' (1975)
Reprinted in:
M. Harloe (ed.) (1977), *Captive Cities: Studies in the Political Economy of
Cities and Regions*, Wiley, London, pp. 123-39. Slightly edited
'Communication on Recent Comments by Professor Carter', *Professional Geo-
grapher*, 29(4), November, 405-7. Carter's comments appeared in *Professional
Geographer*, 29(1), February 1977, 101-2

1978

'On Planning the Ideology of Planning' in R.W. Burchell and G. Sternlieb (eds),

*Planning Theory in the 1980s: a Search for Future Directions*, Center for Urban Policy Research, Rutgers University, New Brunswick, pp. 213-33

'The Urban Process under Capitalism: a Framework for Analysis', *International Journal of Urban and Regional Research*, 2(1), March, 101-31. Originally presented to the 'Zone' workshop on 'Marxism, Imperialism and the Spatial Analysis of Peripheral Capitalism' in May 1977, in Amsterdam
Reprinted in:
M. Dear and A.J. Scott (eds) (1981), *Urbanization and Urban Planning in Capitalist Society*, Methuen, London, pp. 91-121

'On Countering the Marxian Myth – Chicago Style', *Comparative Urban Research*, 6(2/3), 28-45. A contribution to a symposium on Marx and the city

'The Subversion of Tenure on the Homewood Campus', *Baltimore Sun*

'Karl Marx and the Boundaries of Academic Freedom', *Baltimore Sun*, 14 May, section K, p. 2
Reprinted in a slightly shorter version, as 'On Repressive Tolerance', *The Progressive*, October 1978, 30-1

## 1979

'Monument and Myth', *Annals of the Association of American Geographers*, 69(3), September, 362-81

## 1980

'Nicaragua Rebuilds', *The Progressive*, May, 42-6. With B. Koeppel

'Tragedy in El Salvador', *The Progressive*, May, 44-5. With B. Koeppel

## 1981

'Rent Control and a Fair Return' in J.I. Gilderbloom *et al.*, *Rent Control: a Source Book*, Foundation for National Progress, Santa Barbara, pp. 80-2. Originally published in the *Baltimore Sun*, 20 September 1979

# SELECT BIBLIOGRAPHY*

Aay, H. (1978) 'Conceptual Change and the Growth of Geographic Knowledge: a Critical Appraisal of the Historiography of Geography', unpublished PhD dissertation, Department of Geography, Clark University, Worcester

Abel, T. (1948) 'The Operation Called Verstehen', *American Journal of Sociology*, 54, 211-18

Ackerman, E.A. (1963) 'Where is a Research Frontier?', *Annals, AAG*, 53, 429-40

——, Berry, B.J.L., Bryson, R.A., Cohen, S.B., Taaffe, E.J., Thomas Jr, W.L. and Wolman, M.G. (1965) *The Science of Geography*, National Academy of Sciences, National Research Council, Publication No. 1277, Washington, DC

Ackoff, R.L. (1962) *Scientific Method: Optimizing Applied Research Decisions*, Wiley, New York

Albaum, M. (ed.) (1973) *Geography and Contemporary Issues: Studies of Relevant Problems*, Wiley, New York

Allen, R.G.D. (1949) *Statistics for Economists*, Hutchinson, London

Alonso, W. (1964) *Location and Land Use*, Harvard University Press, Cambridge, Massachusetts

Amedeo, D. (1971) 'Review of D. Harvey (1969), *Explanation in Geography*', *Geographical Review*, 61, 147-9

—— and Golledge, R.G. (1975) *An Introduction to Scientific Reasoning in Geography*, Wiley, New York

Amin, S. (1973) *Accumulation on a World Scale*, Monthly Review Press, New York

Anderson, J. (1980) 'Towards a Materialist Conception of Geography', *Geoforum*, 11, 171-8

Bach, W. (1972) *Atmospheric Pollution*, McGraw-Hill, New York

Ballabon, M.G. (1957) 'Putting the "Economic" into Economic Geography', *Economic Geography*, 33, 217-23

Baran, P. (1957) *The Political Economy of Growth*, Monthly Review Press, New York

—— and Sweezy, P. (1966) *Monopoly Capital*, Penguin, Harmondsworth

Barnbrock, J. (1974) 'Prolegomenon to a Methodological Debate on Location Theory: the Case of von Thunen', *Antipode*, 6(1), 59-66

Barraclough, G. (1957) *History in a Changing World*, Blackwell, Oxford

Bennett, R.J. and Chorley, R.J. (1978) *Environmental Systems: Philosophy, Analysis and Control*, Methuen, London

Benton, T. (1977) *Philosophical Foundations of the Three Sociologies*, Routledge and Kegan Paul, London

Berdoulay, V. (1976) 'French Possibilism as a Form of Neo-Kantian Philosophy', *Proceedings, AAG*, 8, 176-9

Berry, B.J.L. (1964a) 'Approaches to Regional Analysis: a Synthesis', *Annals, AAG*, 54, 2-11

—— (1964b) 'Cities as Systems within Systems of Cities', *Papers of the Regional Science Association*, 13, 147-63

—— (1972a) 'More on Relevance and Policy Analysis', *Area*, 4, 77-80

—— (1972b) '"Revolutionary and Counter-revolutionary Theory in Geography" – a Ghetto Commentary', *Antipode*, 4(2), 31-3

*AAG – Association of American Geographers. IBG – Institute of British Geographers

185

—— (1973) 'A Paradigm for Modern Geography' in R.J. Chorley (ed.), *Directions in Geography*, Methuen, London, pp. 3-21

—— (1974a) 'Review of D. Harvey (1973), *Social Justice and the City*', *Antipode*, 6(2), 142-5

—— (1974b) 'Review of H.M. Rose (ed.) (1972), *The Geography of the Ghetto: Perceptions, Problems and Alternatives*', *Annals, AAG*, 64, 342-5

—— (1975) 'Short-term Housing Cycles in a Dualistic Metropolis' in G. Gappert and H.M. Rose (eds), *The Social Economy of Cities*, Urban Affairs Annual, No. 9, Sage Publications, Beverley Hills, pp. 165-82

—— (1978) 'Introduction: a Kuhnian Perspective' in B.J.L. Berry (ed.), *The Nature of Change in Geographical Ideas*, Perspectives in Geography, vol. 3, Northern Illinois University Press, De Kalb, pp. vii-x

— (1980a) 'Review of S. Gale and G. Olsson (eds) (1979), *Philosophy in Geography*', *Canadian Geographer*, 24, 197-8

—— (1980b) 'Creating Future Geographies', *Annals, AAG*, 70, 449-58

—— and Garrison, W.L. (1958a) 'The Functional Bases of the Central Place Hierarchy', *Economic Geography*, 34, 145-54

—— and —— (1958b) 'Recent Developments in Central Place Theory', *Papers and Proceedings of the Regional Science Association*, 4, 107-20

—— and Pred, A. (1961) *Central Place Studies: a Bibliography of Theory and Applications*, Regional Science Research Institute, Bibliography Series, No. 1, Philadelphia

Birch, J.W. (1971) 'Review of D. Harvey (1969), *Explanation in Geography*', *Geography*, 56, 262-3

Bird, J. (1977) 'Methodology and Philosophy', *Progress in Human Geography*, 1, 104-10

—— (1978) 'Methodology and Philosophy', *Progress in Human Geography*, 2, 133-40

—— (1979) 'Methodology and Philosophy', *Progress in Human Geography*, 3, 117-25

Blaut, J.M. (1975) 'Imperialism, the Marxist Theory and its Evolution', *Antipode*, 7(1), 1-19

—— (1980) 'A Radical Critique of Cultural Geography', *Antipode*, 12(2), 25-9

Blowers, A.T. (1972) 'Bleeding Hearts and Open Values', *Area*, 4, 290-2

—— (1974) 'Relevance, Research and the Political Process', *Area*, 6, 32-6

Boddy, M. (ed.) (1976) 'Urban Political Economy', *Antipode*, 8(1)

Braithwaite, R.B. (1960) *Scientific Explanation: a Study of the Function of Theory, Probability and Law in Science*, Harper Torchbooks, New York

Breitbart, M. (1975) 'Impressions of an Anarchist Landscape', *Antipode*, 7(2), 44-9

—— (1979) 'Anarchism: the Spanish Experience', *Antipode*, 10(3)/11(1), 60-70

Brodbeck, M. (1959) 'Models, Meaning and Theories' in L. Gross (ed.), *Symposium on Sociological Theory*, Harper and Row, New York, pp. 373-403

Brookfield, H.C. (1964) 'Questions on the Human Frontiers of Geography', *Economic Geography*, 40, 283-303

—— (1973) 'Introduction: Explaining or Understanding? The Study of Adaptation and Change' in H.C. Brookfield (ed.), *The Pacific in Transition: Geographical Perspectives on Adaptation and Change*, Edward Arnold, London, pp. 3-23

Brown, H.I. (1977) *Perception, Theory and Commitment: the New Philosophy of Science*, Precedent, Chicago

Bruegel, I. (1975) 'The Marxist Theory of Rent and the Contemporary City: a Critique of Harvey' in *Political Economy and the Housing Question*, Political Economy of Housing Workshop, London, pp. 34-46

Brunhes, J. (1912) *La Géographie Humaine*, Presses Universitaires de France, Paris
Brunn, S.D. (1974) 'Review of D. Harvey (1973), *Social Justice and the City*', *Professional Geographer*, 26, 342-3
Buchdahl, G. (1965) 'A Revolution in Historiography of Science', *History of Science*, 4, 55-69
Bunge, W. (1962) *Theoretical Geography*, Lund Studies in Geography, Series C, No. 1, Royal University of Lund, Sweden
—— (1969) 'The First Years of the Detroit Geographical Expedition: a Personal Report', *Field Notes*, 1, 1-9
—— (1971) *Fitzgerald: Geography of a Revolution*, Schenkman, Cambridge, Massachusetts
Burchell, R.W. and Sternlieb, G. (eds) (1978) *Planning Theory in the 1980s: a Search for Future Directions*, Center for Urban Policy Research, Rutgers University, New Brunswick
Burgess, R. (1976) *Marxism and Geography*, University College, London, Department of Geography, Occasional Paper No. 30
—— (1977) 'Self-help Housing: a New Imperialist Strategy? A Critique of the Turner School', *Antipode*, 9(2), 50-9
Burton, I. (1963) 'The Quantitative Revolution and Theoretical Geography', *Canadian Geographer*, 7, 151-62
Butterfield, G.R. (1977) 'Science and Explanation? The Scientific Method in Geography' in R. Lee (ed.), *Change and Tradition: Geography's New Frontiers*, Department of Geography, Queen Mary College, University of London
Buttimer, A. (1971) *Society and Milieu in the French Geographic Tradition*, Rand McNally, Chicago
—— (1976) 'Grasping the Dynamism of Lifeworld', *Annals, AAG*, 66, 277-92
—— (1977) 'Comment in Reply', *Annals, AAG*, 67, 180-3
—— (1981) 'On People, Paradigms, and "Progress" in Geography' in D.R. Stoddart (ed.), *Geography, Ideology and Social Concern*, Basil Blackwell, Oxford, pp. 81-98
Campbell, J.S. (1972) 'Libertarian Reactions to a Marxist View: Comment on David Harvey', *Antipode*, 4(2), 21-5
Caponigri, A.R. (1971) *A History of Western Philosophy*, vol. 5 – *Philosophy from the Age of Positivism to the Age of Analysis*, University of Notre Dame Press, Indiana
Carey, G.W. (1975) 'Review of D. Harvey (1973), *Social Justice and the City*', *Geographical Review*, 65, 421-3
Carnap, R. (1936) 'Testability and Meaning, I', *Philosophy of Science*, 3, 419-71
—— (1937) 'Testability and Meaning, II', *Philosophy of Science*, 4, 1-40
—— (1942) *Introduction to Semantics*, Harvard University Press, Cambridge, Massachusetts
—— (1956) 'The Methodological Character of Theoretical Concepts', *Minnesota Studies in the Philosophy of Science*, 1, 38-76
—— (1958) *Introduction to Symbolic Language and its Applications*, Dover, New York
Carter, G.F. (1977) 'A Geographical Society Should be a Geographical Society', *Professional Geographer*, 29, 101-2
Castells, M. (1970) 'Structures Sociales et Processus d'Urbanization', *Annales, Economies, Sociétés, Civilization*, 25, 1155-99
—— (1972) *La Question Urbaine*, Maspero, Paris
—— (1977) *The Urban Question: a Marxist Approach*, Edward Arnold, London
Chapman, G.P. (1977) *Human and Environmental Systems: a Geographer's Appraisal*, Academic Press, London
Chatterjee, L. (1973) 'Real Estate Investment and Deterioration of Housing in

Baltimore', unpublished PhD dissertation, Department of Geography and Environmental Engineering, Johns Hopkins University, Baltimore

——, Harvey, D. and Klugman, L. (1974) *FHA Policies and the Baltimore City Housing Market*, Baltimore Urban Observatory, Baltimore

Chisholm, M. (1967) 'General Systems Theory and Geography', *Transactions, IBG*, 24, 42-52

—— (1971) 'Geography and the Question of "Relevance"', *Area*, 3, 65-8

—— (1975) *Human Geography: Evolution or Revolution?*, Penguin, Harmondsworth

——, Frey, A.E. and Haggett, P. (eds) (1971) *Regional Forecasting: Proceedings of the Twenty-second Symposium of the Colston Research Society*, Butterworths, London

Chorley, R.J. (1962) *Geomorphology and General Systems Theory*, Geological Survey Professional Paper 500-B, US Government Printing Office, Washington DC

—— (1964) 'Geography and Analogue Theory', *Annals, AAG*, 54, 127-37

—— and Haggett, P. (1965) 'Trend Surface Mapping in Geographic Research', *Transactions, IBG*, 37, 47-67

—— and —— (eds) (1967) *Models in Geography*, Methuen, London

Cliff, A.D. and Ord, J.K. (1973) *Spatial Autocorrelation*, Pion, London

Colby, C.C. (1936) 'Changing Currents of Geographic Thought in America', *Annals, AAG*, 26, 1-37

Cole, J.P. and King, C.A.M. (1968) *Quantitative Geography: Techniques and Theories in Geography*, Wiley, London

Collingwood, R.G. (1956) *The Idea of History*, Oxford University Press, New York

Cooper, M. (1979) 'The Urban Experience of Aborigines: a Structural Analysis', *Antipode*, 11(3), 67-76

Cox, K.R. (1972) *Man, Location, and Behavior: an Introduction to Human Geography*, Wiley, New York

—— (1973) *Conflict, Power, and Politics in the City: a Geographic View*, McGraw-Hill, New York

—— (1976) 'Review of D. Harvey (1973), *Social Justice and the City*', *Geographical Analysis*, 8, 333-7

—— and Golledge, R.G. (eds) (1969a) *Behavioral Problems in Geography: a Symposium*, Northwestern University Studies in Geography, 17, Evanston, Illinois

—— and —— (1969b) 'Editorial Introduction: Behavioral Models in Geography' in K.R. Cox and R.G. Golledge (eds), *Behavioral Problems in Geography: a Symposium*, Northwestern University Studies in Geography, 17, Evanston, Illinois, pp. 1-13

—— and —— (eds) (1981) *Behavioural Problems in Geography Revisited*, Methuen, London

Cumberland, K.B. (1950) 'The Geographer's Point of View', Inaugural Lecture, delivered 2 April 1946, Auckland University College, Auckland

Curry, L. (1967a) 'Quantitative Geography, 1967', *Canadian Geographer*, 11, 265-79

—— (1967b) 'Central Places in the Random Spatial Economy', *Journal of Regional Science*, 7 (supplement), 217-38

Dacey, M.F. (1962) 'Analysis of Central Place and Point Patterns by a Nearest Neighbour Method' in K. Norberg (ed.), *IGU Symposium in Urban Geography*, Lund

—— (1964) 'Modified Poisson Probability Law for Point Pattern More Regular than Random', *Annals, AAG*, 54, 559-65

Dahrendorf, R. (1959) *Class and Class Conflict in Industrial Society*, Stanford University Press, Stanford, California

Darby, H.C. (1946) 'Theory and Practice of Geography', Inaugural Lecture, University of Liverpool

―― (1953) 'On the Relations of Geography and History', *Transactions and Papers, IBG*, 19, 1-11

De Fleury, H.R. (1903) *Historique de la Basilique de Sacré-Coeur*, vol. 1, Bibliothèque Nationale, Paris

Dickinson, R.E. (1951) *The West European City*, Routledge and Kegan Paul, London

―― (1969) *The Makers of Modern Geography*, Praeger, New York

Doherty, J. (1973) 'Race, Class and Residential Segregation in Britain', *Antipode*, 5(3), 45-51

Doornkamp, J.C. and Warren, K. (1980) 'Geography in the United Kingdom, 1976-1980', *Geographical Journal*, 146, 94-110

Dooyeweerd, H. (1953) *A New Critique of Theoretical Thought*, vol. 1, *The Necessary Presuppositions of Philosophy*, Presbyterian and Reformed, Philadelphia

―― (1960) *In the Twilight of Western Thought*, Craig Press, Nutley, New Jersey

Douglas, J.B. (1955) 'Fitting the Neyman Type A (Two-parameter) Distribution', *Biometrics*, 11, 149-73

Downs, R.M. and Stea, D. (1977) *Maps in Minds: Reflections on Cognitive Mapping*, Harper and Row, New York

Dunbar, G. (1979) 'Élisée Reclus, geographer and anarchist', *Antipode*, 10(3)/ 11(1), 16-21

Duncan, C. (1966) 'Exploring the Frontiers' in *Seven Inaugural Lectures*, University of Waikato, Hamilton, pp. 85-105

―― (1972) 'The Man-land Interface – a Contemporary Theme in Applied Geography', *Proceedings of the Seventh Geography Conference*, New Zealand Geographical Society, pp. 137-43

Duncan, J.S. and Duncan, N.G. (1976) 'Social Worlds, Status Passage, and Environmental Perspectives' in G.T. Moore and R.G. Golledge (eds), *Environmental Knowing: Theories, Research and Methods*, Dowden, Hutchinson and Ross, Stroudsberg, Pennsylvania, pp. 206-13

―― and Ley, D. (1982) 'Structural Marxism and Human Geography: a Critical Reassessment', *Annals, AAG*, 72, 30-59

Duncan, O.D., Cuzzort, R.P. and Duncan, B. (1961) *Statistical Geography*, Free Press, London

Edelson, N. and Eliot Hurst, M.E. (1976) 'Towards an Outline of a Marxian Perspective: a Comment on N.E.P.', *Antipode*, 8(2), pp. 74-80

Engels, F. (1935) *The Housing Question*, International Publishers, New York

―― (1946) *Dialectics of Nature*, Lawrence and Wishart, London

―― (1958) *The Condition of the Working Class in England*, Stanford University Press, Stanford, California

Entrikin, J.N. (1976) 'Contemporary Humanism in Geography', *Annals, AAG*, 66, 615-32

―― (1977) 'Geography's Spatial Perspective and the Philosophy of Ernst Cassirer', *Canadian Geographer*, 21, 209-22

Ernst, B. (1976) *The Magic Mirror of M.C. Escher*, Ballantine Books, New York

Ettorre, E.M. (1978) 'Women, Urban Social Movements and the Lesbian Ghetto', *International Journal of Urban and Regional Research*, 2, 499-520

Eyles, J. (1973) 'Geography and Relevance', *Area*, 5, 158-60

Feyerabend, P. (1963) 'How to be a Good Empiricist – a Plea for Tolerance in Matters Epistemological' in B. Baumrin (ed.), *Philosophy of Science: the Delaware Seminar*, vol. 2, Interscience, New York, pp. 3-40

Finch, V.C. (1939) 'Geographical Science and Social Philosophy', *Annals, AAG*, 29, 1-28

Folke, S. (1972) 'Why a Radical Geography Must Be Marxist', *Antipode*, 4(2), 13-18

Foraie, J. and Dear, M. (1978) 'The Politics of Discontent among Canadian Indians', *Antipode*, 10(1), 34-45

Foster, J. (1974) *Class Struggle and the Industrial Revolution − Early Industrial Capitalism in Three English Towns*, Weidenfeld and Nicolson, London

Frank, A.G. (1969) *Capitalism and Underdevelopment in Latin America*, Monthly Review Press, New York

Freeman, D. (1966) 'Social Anthropology and the Scientific Study of Human Behaviour', *Man*, n.s.1, 330-42

Fried, M. (1967) *The Evolution of Political Society*, Random House, New York

Gale, S. (1972a) 'On the Heterodoxy of Explanation: a Review of David Harvey's *Explanation in Geography*', *Geographical Analysis*, 4, 285-322

—— (1972b) 'Inexactness, Fuzzy Sets and the Foundations of Behavioral Geography', *Geographical Analysis*, 4, 337-50

—— and Olsson, G. (eds) (1979a) *Philosophy in Geography*, Reidel, Dordrecht

—— and —— (1979b) 'Introduction' in S. Gale and G. Olsson (eds), *Philosophy in Geography*, Reidel, Dordrecht, pp. ix-xxi

Galois, B. (1976) 'Ideology and the Idea of Nature. The Case of Peter Kropotkin', *Antipode*, 8(3), 1-16

Gappert, G. and Rose, H.M. (eds) (1975) *The Social Economy of Cities*, Urban Affairs Annual No. 9, Sage Publications, Beverley Hills

Garrison, W.L. (1956) 'Applicability of Statistical Inference to Geographical Research', *Geographical Review*, 46, 427-9

—— (1959a) 'Spatial Structure of the Economy: I', *Annals, AAG*, 49, 232-9

—— (1959b) 'Spatial Structure of the Economy: II', *Annals, AAG*, 49, 471-82

—— (1960) 'Spatial Structure of the Economy: III', *Annals, AAG*, 50, 357-73

—— and Marble, D.F. (1957) 'The Spatial Structure of Agricultural Activities', *Annals, AAG*, 47, 137-44

Getis, A. (1972) 'Other Revolutionary Paradigms: Comments on Harvey's Paper', *Antipode,* 4(2), 33-6

Giddens, A. (1973) *The Class Structure of the Advanced Societies*, Hutchinson, London

Gillispie, C.C. (1973) 'Koyré, Alexandre' in *Dictionary of Scientific Biography*, vol. 7, Scribner's, New York, pp. 482-90

Godelier, M. (1972) *Rationality and Irrationality in Economics*, New Left Books, London

Golledge, R.G. and Amedeo, D. (1968) 'On Laws in Geography', *Annals, AAG*, 58, 760-74

Gould, P.R. (1963) 'Man against his Environment: a Game Theoretic Framework', *Annals, AAG*, 53, 290-7

—— (1966) *On Mental Maps*, Michigan Inter-university Community of Mathematical Geographers, Discussion Paper No. 9

—— and White, R. (1974) *Mental Maps*, Penguin, Harmondsworth

Gouldner, A.W. (1980) *The Two Marxisms: Contradictions and Anomalies in the Development of Theory*, Seabury, New York

Gramsci, A. (1971) *Selections from the Prison Notebooks*, Lawrence and Wishart, London

Gray, F. (1975) 'Non-explanation in Urban Geography', *Area*, 7, 228-35

Gregory, D. (1978) *Ideology, Science and Human Geography*, Hutchinson, London

Gregory, S. (1963) *Statistical Methods and the Geographer*, Longmans, London

—— (1976) 'On Geographical Myths and Statistical Fables', *Transactions, IBG*, 1, 385-400

Greig-Smith, P. (1964) *Quantitative Plant Ecology*, 2nd edn, Butterworths, London

Guelke, L. (1971) 'Problems of Scientific Explanation in Geography', *Canadian Geographer*, 15, 38-53

—— (1974) 'An Idealist Alternative in Human Geography', *Annals, AAG*, 64, 193-202

—— (1976) 'Frontier Settlement in Early Dutch South Africa', *Annals, AAG*, 66, 25-42

—— (1977a) 'The Role of Laws in Human Geography', *Progress in Human Geography* 1, 376-86

—— (1977b) 'Regional Geography', *Professional Geographer*, 29, 1-7

—— (1978) 'Geography and Logical Positivism' in D.T. Herbert and R.J. Johnston (eds), *Geography and the Urban Environment: Progress in Research and Applications*, vol. 1, Wiley, London, pp. 35-61

—— (1981) 'Idealism' in M.E. Harvey and B.P. Holly (eds), *Themes in Geographic Thought*, Croom Helm, London, pp. 133-47

Hagerstrand, T. (1953) *Innovationsforloppet ur Korologisk Synpunkt*, Gleerup, Lund

—— (1967) 'The Computer and Geography', *Transactions, IBG*, 42, 1-19

Haggett, P. (1965) *Locational Analysis in Human Geography*, Edward Arnold, London

—— and Chorley, R.J. (1967) 'Models, Paradigms and the New Geography' in R.J. Chorley and P. Haggett (eds), *Models in Geography*, Methuen, London, pp. 19-41

——, Cliff, A.D. and Frey, A. (1977) *Locational Analysis in Human Geography*, Edward Arnold, London

Hall, P. (1973) 'Review of D. Harvey (1973), *Social Justice and the City* and T. Blair (1973), *The Poverty of Planning: Crisis in the Urban Environment*', *New Society*, 25, 567, 408-9

Halvorson, P. and Stave, B.M. (1978) 'A Conversation with Brian J.L. Berry', *Journal of Urban History*, 4, 209-38

Hanson, N.R. (1965) *Patterns of Discovery*, Cambridge University Press, Cambridge

Harloe, M. (ed.) (1977) *Captive Cities: Studies in the Political Economy of Cities and Regions*, Wiley, London

—— (1979) 'Editorial', *International Journal of Urban and Regional Research*, 3, 1-2

Harries, K.D. (1974) *The Geography of Crime and Justice*, McGraw-Hill, New York

Harrison, R.T. and Livingstone, D.N. (1980) 'Philosophy and Problems in Human Geography: a Presuppositional Approach', *Area*, 12, 25-31

Hartshorne, R. (1939) *The Nature of Geography: a Critical Survey of Current Thought in the Light of the Past*, Association of American Geographers, Lancaster, Pennsylvania

—— (1948) 'On the Mores of Methodological Discussion', *Annals, AAG*, 38, 113-25

—— (1958) 'The Concept of Geography as a Science of Space, from Kant to Humboldt to Hettner', *Annals, AAG*, 48, 97-108

—— (1959) *Perspective on the Nature of Geography*, Rand McNally, Chicago

—— (1972) 'Review of J.A. May (1970), *Kant's Concept of Geography*', *Canadian Geographer*, 16, 77-9

Harvey, D. (1961) 'Aspects of Agricultural and Rural Change in Kent, 1800-1900',

PhD dissertation, Department of Geography, University of Cambridge
—— (1963) 'Locational Change in the Kentish Hop Industry and the Analysis of Land Use Patterns', *Transactions and Papers, IBG*, 33, 123-44
—— (1965) 'Simulation Models' in G. Olsson and O. Warneryd (eds), *Meddelande fran ett Symposium i Teoretisk Samhallsgeografi*, Forskningsrapporter fran Kulturgeografiska Institutionen, Uppsala Universitet, No. 1, pp. 47-8
—— (1966a) 'Theoretical Concepts and the Analysis of Agricultural Land-use Patterns', *Annals, AAG*, 56, 361-74
—— (1966b) 'Geographical Processes and the Analysis of Point Patterns: Testing Models of Diffusion by Quadrat Sampling', *Transactions, IBG*, 40, 81-95
—— (1967a) 'Models of the Evolution of Spatial Patterns in Human Geography' in R.J. Chorley and P. Haggett (eds), *Models in Geography*, Methuen, London, pp. 549-608
—— (1967b) *Behavioural Postulates and the Construction of Theory in Human Geography*, University of Bristol, Department of Geography, Seminar Paper, Series A, No. 6
—— (1967c) 'Editorial Introduction: the Problem of Theory Construction in Geography', *Journal of Regional Science*, 7 (supplement), 211-16
—— (1968a) 'Some Methodological Problems in the Use of the Neyman Type A and the Negative Binomial Probability Distributions for the Analysis of Spatial Point Patterns', *Transactions, IBG*, 44, 85-95
—— (1968b) 'Pattern, Process, and the Scale Problem in Geographical Research', *Transactions, IBG*, 45, 71-8
—— (1969a) *Explanation in Geography*, Edward Arnold, London
—— (1969b) 'Conceptual and Measurement Problems in the Cognitive-behavioral Approach to Location Theory' in K.R. Cox and R.G. Golledge (eds), *Behavioral Problems in Geography: a Symposium*, Northwestern University Studies in Geography, 17, Evanston, Illinois, pp. 35-68
—— (1969c) 'Review of A. Pred (1967), *Behavior and Location: Foundations for a Geographic and Dynamic Location Theory*, Part I', *Geographic Review*, 59, 312-14
—— (1970a) 'Social Processes and Spatial Form: an Analysis of the Conceptual Problems of Urban Planning', *Papers of the Regional Science Association*, 25, 47-69
—— (1970b) 'Behavioural Postulates and the Construction of Theory in Human Geography', *Geographia Polonica*, 18, 27-45
—— (1970c) 'Locational Change in the Kentish Hop Industry and the Analysis of Land Use Patterns: Supplementary Note' in A.R.H. Baker, J.D. Hamshere and J. Langton (eds), *Geographical Interpretations of Historical Sources: Readings in Historical Geography*, David and Charles, Newton Abbot, pp. 264-5
—— (1971) 'Social Processes, Spatial Form and the Redistribution of Real Income in an Urban System' in M. Chisholm, A.E. Frey and P. Haggett (eds), *Regional Forecasting: Proceedings of the Twenty-second Symposium of the Colston Research Society*, Butterworths, London, pp. 267-300
—— (1972a) 'The Role of Theory' in N. Graves (ed.), *New Movements in the Study and Teaching of Geography*, Maurice Temple Smith, London, pp. 29-41
—— (1972b) 'Social Justice and Spatial Systems' in R. Peet (ed.), *Geographical Perspectives on American Poverty*, Antipode Monographs in Social Geography No. 1, Worcester, Massachusetts, pp. 87-106
—— (1972c) 'Revolutionary and Counter-revolutionary Theory in Geography and the Problem of Ghetto Formation' in H.M. Rose (ed.), *Geography of the Ghetto: Perceptions, Problems, and Alternatives*, Perspectives in Geography, vol. 2, Northern Illinois University Press, De Kalb, pp. 1-25
—— (1972d) 'Revolutionary and Counter-revolutionary Theory in Geography

and the Problem of Ghetto Formation', *Antipode*, 4(2), 1-13

—— (1972e) 'A Commentary on the Comments', *Antipode*, 4(2), 36-41

—— (1972f) 'On Obfuscation in Geography: a Comment on Gale's Heterodoxy', *Geographical Analysis*, 4, 323-30

—— (1972g) *Society, the City, and the Space-economy of Urbanism*, Association of American Geographers, Commission on College Geography, Resource Paper No. 18, Washington, DC

—— (1972h) 'Review of P. Wheatley (1971), *The Pivot of the Four Quarters: a Preliminary Enquiry into the Origins and Character of the Ancient Chinese City*', *Annals, AAG*, 62, 509-13

—— (1973a) *Social Justice and the City*, Edward Arnold, London

—— (1973b) *A Question of Method for a Matter of Survival*, University of Reading, Department of Geography, Geographical Paper No. 23

—— (1973c) 'A Comment on Morrill's Reply', *Antipode*, 5(2), 86-8

—— (1974a) 'What Kind of Geography for What Kind of Public Policy?', *Transactions, IBG*, 63, 18-24

—— (1974b) 'Class-monopoly Rent, Finance Capital and the Urban Revolution', *Regional Studies*, 8, 239-55

—— (1974c) 'Ideology and Population Theory', *International Journal of Health Services*, 4, 515-37

—— (1974d) 'Population, Resources, and the Ideology of Science', *Economic Geography*, 50, 256-77

—— (1975a) 'Class Structure in a Capitalist Society and the Theory of Residential Differentiation' in R. Peel, M. Chisholm and P. Haggett (eds), *Processes in Physical and Human Geography: Bristol Essays*, Heinemann, London, pp. 254-72

—— (1975b) 'Some Remarks on the Political Economy of Urbanism', *Antipode*, 7(1), 54-61

—— (1975c) 'The Political Economy of Urbanization in Advanced Capitalist Countries: the Case of the United States' in G. Gappert and H.M. Rose (eds), *The Social Economy of Cities*, Urban Affairs Annual No. 9, Sage Publications, Beverly Hills, pp. 119-63

—— (1975d) 'The Geography of Capitalist Accumulation: a Reconstruction of the Marxian Theory', *Antipode*, 7(2), 9-21

—— (1975e) 'Review of B.J.L. Berry (1974), *The Human Consequences of Urbanisation*', *Annals, AAG*, 65, 99-108

—— (1975f) 'Review of J. Foster (1974), *Class Struggle and the Industrial Revolution – Early Industrial Capitalism in Three English Towns*', *Journal of Historical Geography*, 1, 109-11

—— (1976a) 'Labor, Capital, and Class Struggle around the Built Environment in Advanced Capitalist Societies', *Politics and Society*, 6, 265-95

—— (1976b) 'The Marxian Theory of the State', *Antipode*, 8(2), 80-9

—— (1977a) 'Government Policies, Financial Institutions and Neighbourhood Change in U.S. Cities' in D.R. Deskins, G. Kish, J.D. Nystuen and G. Olsson (eds), *Geographic Humanism, Analysis and Social Action: Proceedings of Symposia Celebrating a Half Century of Geography at Michigan*, Michigan Geographical Publications No. 17, University of Michigan, pp. 291-320

—— (1977b) 'Government Policies, Financial Institutions and Neighbourhood Change in United States Cities' in M. Harloe (ed.), *Captive Cities: Studies in the Political Economy of Cities and Regions*, Wiley, London, pp. 123-39

—— (1977c) 'Communication on Recent Comments by Professor Carter', *Professional Geographer*, 29, 405-7

—— (1978a) 'On Planning the Ideology of Planning' in R.W. Burchell and G. Sternlieb (eds), *Planning Theory in the 1980s: a Search for Future Directions*,

Center for Urban Policy Research, Rutgers University, New Brunswick, pp. 213-33
—— (1978b) 'The Urban Process under Capitalism: a Framework for Analysis', *International Journal of Urban and Regional Research*, 2, 101-31
—— (1978c) 'On Countering the Marxian Myth – Chicago Style', *Comparative Urban Research*, 6(2/3), 28-45
—— (1978d) 'The Subversion of Tenure on the Homewood Campus', *Baltimore Sun*
—— (1978e) 'Karl Marx and the Boundaries of Academic Freedom', *Baltimore Sun*, 14 May, K2
—— (1978f) 'On Repressive Tolerance', *The Progressive*, October, 30-1
—— (1979) 'Monument and Myth', *Annals, AAG*, 69, 362-81
—— (1981) 'Rent Control and a Fair Return' in J.I. Gilderbloom *et al.*, *Rent Control: a Source Book*, Foundation for National Progress, Santa Barbara, pp. 80-2
—— and Berry, B.J.L. (1974) 'Discussion', *Antipode*, 6(2), 145-9
—— and Chatterjee, L. (1974) 'Absolute Rent and the Structuring of Space by Governmental and Financial Institutions', *Antipode*, 6(1), 22-36
——, ——, Wolman, M.G., Klugman, L. and Newman, J.S. (1972) *The Housing Market and Code Enforcement in Baltimore*, Baltimore Urban Observatory, Baltimore
Harvey, M.E. and Holly, B.P. (eds) (1981a) *Themes in Geographic Thought*, Croom Helm, London
—— and —— (1981b) 'Paradigm, Philosophy and Geographic Thought' in M.E. Harvey and B.P. Holly (eds), *Themes in Geographic Thought*, Croom Helm, London, pp. 11-37
Hayford, A.M. (1972) 'Comments on Harvey's Paper', *Antipode*, 4(2), 19-21
Hempel, C.G. (1959) 'The Logic of Functional Analysis' in L. Gross (ed.), *Symposium on Sociological Theory*, Row, Peterson, Illinois, 271-307
—— (1965) *Aspects of Scientific Explanation and Other Essays in the Philosophy of Science*, Free Press, New York
—— and Oppenheim, P. (1948) 'Studies in the Logic of Explanation', *Philosophy of Science*, 15, 135-75
Hill, M.R. (1981) 'Positivism: a "Hidden" Philosophy in Geography' in M.E. Harvey and B.P. Holly (eds), *Themes in Geographic Thought*, Croom Helm, London, pp. 38-60
Hindness, B. (1979) 'Review of D. Gregory (1978), *Ideology, Science and Human Geography*', *Environment and Planning A*, 11, 350-2
Hoch, C. (1979) 'Social Structure and Suburban Spatio-political Conflicts in the United States', *Antipode*, 11(3), 44-55
Holt-Jensen, A. (1980) *Geography: its History and Concepts*, Harper and Row, London
Hoselitz, B.F. (1960) *Sociological Aspects of Economic Growth*, Free Press, New York
Huberman, L. and Sweezy, P.M. (1969) *Socialism in Cuba*, Monthly Review Press, New York
Huggett, R.J. (1980) *Systems Analysis in Geography*, Clarendon Press, Oxford
Isard, W. (1956a) *Location and Space Economy*, Wiley, New York
—— (1956b) 'Regional Science, the Concept of Region and Regional Structure', *Papers and Proceedings of the Regional Science Association*, 2, 13-39
—— (1960) *Methods of Regional Analysis: an Introduction in Regional Science*, Wiley, New York
—— (1975) *An Introduction to Regional Science*, Prentice-Hall, Englewood Cliffs
—— and Dacey, M.F. (1962) 'On the Projection of Individual Behavior in Regional Analysis', *Journal of Regional Science*, 4, 1-32 and 51-83

James, P.E. (1952) 'Towards a Further Understanding of the Regional Concept', *Annals, AAG*, 42, 195-222
—— (1972) *All Possible Worlds: a History of Geographical Ideas*, Odyssey, Indianapolis
Joergensen, J. (1951) *The Development of Logical Empiricism*, University of Chicago Press, Chicago
Johnson, H.G. (1971) 'The Keynesian Revolution and the Monetarist Counter-revolution', *American Economic Review*, 61(2), 1-14
Johnston, R.J. (1972) 'Continually Changing Human Geography: a Review of Some Recent Literature', *New Zealand Geographer*, 28, 78-96
—— (1974) 'Continually Changing Human Geography Revisited: David Harvey: *Social Justice and the City*', *New Zealand Geographer*, 30, 180-92
—— (1977) 'Urban Geography: City Structures', *Progress in Human Geography*, 1, 118-29
—— (1979) *Geography and Geographers: Anglo-American Human Geography since 1945*, Edward Arnold, London
Kao, R.C. (1963) 'The Use of Computers in the Processing and Analysis of Geographic Information', *Geographical Review*, 53, 530-47
Kaplan, A. (1964) *The Conduct of Inquiry*, Chandler, San Francisco
Kates, R.W. (1962) *Hazard and Choice Perception in Flood Plain Management*, University of Chicago, Department of Geography, Research Paper No. 78
Keat, R. and Urry, J. (1975) *Social Theory as Science*, Routledge and Kegan Paul, London
Kennedy, B.A. (1970) 'Review of D. Harvey (1969), *Explanation in Geography*', *British Journal for the Philosophy of Science*, 21, 401-2
King, L.J. (1976) 'Alternatives to a Positive Economic Geography', *Annals, AAG*, 66, 293-308
Kirschmann, M.J. (1981) 'Winning Rent Control in a Working Class City' in J.I. Gilderbloom *et al.*, *Rent Control: a Source Book*, Foundation for National Progress, Santa Barbara, pp. 192-6
Klugman, L. (1974) 'The F.H.A. and Home-ownership in the Baltimore Housing Market (1963-1972)', PhD dissertation, Department of Geography, Clark University, Worcester
Knox, P.L. (1975) *Social Well-being: a Spatial Perspective*, Oxford University Press, London
—— and Cottam, M.B. (1981) 'A Welfare Approach to Rural Geography: Contrasting Perspectives on the Quality of Highland Life', *Transactions, IBG*, 6, 433-50
Koeppel, B. and Harvey, D. (1980a) 'Nicaragua Rebuilds', *The Progressive*, May, 42-6
—— and —— (1980b) 'Tragedy in El Salvador', *The Progressive*, May, 44-5
Kolakowski, L. (1975) *Husserl and the Search for Certitude*, Yale University Press, New Haven
Kuhn, T.S. (1957) *The Copernican Revolution: Planetary Astronomy in the Development of Western Thought*, Harvard University Press, Cambridge, Massachusetts
—— (1962) *The Structure of Scientific Revolutions*, University of Chicago Press, Chicago
—— (1963) 'The Function of Dogma in Scientific Research' in A.C. Crombie (ed.), *Scientific Change*, Heinemann, London, pp. 347-69
—— (1970a) 'Logic of Discovery or Psychology of Research?' in I. Lakatos and A. Musgrave (eds), *Criticism and the Growth of Knowledge*, Cambridge University Press, London, pp. 1-23
—— (1970b) 'Reflections on my Critics' in I. Lakatos and A. Musgrave (eds),

*Criticism and the Growth of Knowledge*, Cambridge University Press, London, pp. 231-78

—— (1970c) *The Structure of Scientific Revolutions*, 2nd edn, enlarged, University of Chicago Press, Chicago

—— (1970d) 'Alexandre Koyré and the History of Science', *Encounter*, 24(1), 67-9

—— (1974) 'Second Thoughts on Paradigms' in F. Suppe (ed.), *The Structure of Scientific Theories*, University of Illinois Press, Urbana, pp. 459-99

Laclau, E. (1975) 'The Specificity of the Political: Around the Poulantzas-Miliband Debate', *Economy and Society*, 5, 87-110

Lakatos, I. (1970) 'Falsification and the Methodology of Scientific Research Programmes' in I. Lakatos and A. Musgrave (eds), *Criticism and the Growth of Knowledge*, Cambridge University Press, London, pp. 91-196

—— (1971) 'History of Science and its Rational Reconstruction', *Boston Studies in the Philosophy of Science*, 8, 91-136

Le Gates, R.T. and Murphy, K. (1981) 'Austerity, Shelter, and Social Conflict in the United States', *International Journal of Urban and Regional Research*, 5, 255-75

Lee, D.R. (1974) 'Existentialism in Geographic Education', *Journal of Geography*, 73, 13-19

Lefebvre, H. (1970) *La Revolution Urbaine*, Gallimard, Paris

—— (1972) *La Pensée Marxiste et la Ville*, Casterman, Paris

Leighly, J. (1979) 'Berkeley: Drifting into Geography in the Twenties', *Annals, AAG*, 69, 4-9

Lenin, V.I. (1927) *Materialism and Empirio-criticism*, International Publishers, New York

—— (1965) *Imperialism, the Highest Stage of Capitalism*, Foreign Languages Press, China

Lewis, P.W. (1965) 'Three Related Problems in the Formulation of Laws in Geography', *Professional Geographer*, 17, 24-7

Ley, D. (1977) 'Social Geography and the Taken-for-granted World', *Transactions, IBG*, 2, 498-512

—— (1978) 'Social Geography and Social Action' in D. Ley and M. Samuels (eds), *Humanistic Geography: Prospects and Problems*, Maaroufa, Chicago, pp. 41-57

—— (1981) 'Behavioural Geography and the Philosophies of Meaning' in K.R. Cox and R.G. Golledge (eds), *Behavioural Problems in Geography Revisited*, Methuen, London, pp. 209-30

—— and Samuels, M. (eds) (1978) *Humanistic Geography: Prospects and Problems*, Maaroufa, Chicago

Livingstone, D.N. (1979) 'Some Methodological Problems in the History of Geographical Thought', *Tijdschrift voor Economische en Sociale Geografie*, 70, 226-31

—— and Harrison, R.T. (1981) 'Immanuel Kant, Subjectivism, and Human Geography', *Transactions, IBG*, 6, 359-74

Lopes, J.R.B. (1978) 'Capitalist Development and Agrarian Structure in Brazil', *International Journal of Urban and Regional Research*, 2, 1-11

Losch, A. (1954) *The Economics of Location*, Yale University Press, New Haven

Losee, J. (1980) *A Historical Introduction to the Philosophy of Science*, 2nd edn, enlarged, Oxford University Press, Oxford

Lukacs, G. (1968) *History and Class Consciousness*, Merlin, London

Lukermann, F. (1961) 'The Role of Theory in Geographical Inquiry', *Professional Geographer*, 13, 1-6

Luxemburg, R. (1951) *The Accumulation of Capital*, Routledge and Kegan Paul, London

McLellan, D. (1971) *The Thought of Karl Marx: an Introduction*, Macmillan, London
—— (1972) *Marx before Marxism*, Penguin, Harmondsworth
Mackinder, H. (1917) *Democratic Ideas and Reality*, Constable, London
Marble, D.F. (1967) *Some Computer Programs for Geographic Research*, Northwestern University, Department of Geography, Evanston, Illinois
Marchand, B. (1974) 'Quantitative Geography: Revolution or Counter-revolution?', *Geoforum*, 17, 15-23
—— (1978) 'A Dialectical Approach in Geography', *Geographical Analysis*, 10, 105-19
—— (1979) 'Dialectics and Geography' in S. Gale and G. Olsson (eds), *Philosophy in Geography*, Reidel, Dordrecht, pp. 237-67
Marx, K. (1963) *Early Writings*, translated and edited by T.B. Bottomore, C.A. Watts, London
—— (1964) *The Economic and Philosophic Manuscripts of 1844*, International Publishers, New York
—— (1967) *Capital*, 3 vols, International Publishers, New York
—— (1970) *A Contribution to the Critique of Political Economy*, International Publishers, New York
—— (1973) *The Grundrisse*, Penguin, Harmondsworth
—— and Engels, F. (1971) *The German Ideology*, International Publishers, New York
—— and —— (1975) *Collected Works*, vol. 3, 1843-1844, Lawrence and Wishart, London
Masser, I. (ed.) (1976) *Theory and Practice in Regional Science*, Pion, London
Massey, D. (1974) 'Review of D. Harvey (1973), *Social Justice and the City*', *Environment and Planning A*, 6, 229-35
Masterman, M. (1970) 'The Nature of a Paradigm' in I. Lakatos and A. Musgrave (eds), *Criticism and the Growth of Knowledge*, Cambridge University Press, London, pp. 59-89
May, J.A. (1970) *Kant's Concept of Geography and its Relation to Recent Geographical Thought*, University of Toronto, Department of Geography, Research Publication No. 4, University of Toronto Press, Toronto
—— (1972) 'A Reply to Professor Hartshorne', *Canadian Geographer*, 16, 79-81
Mercer, D.C. (1977) *Conflict and Consensus in Human Geography*, Monash Publications in Geography No. 17, Melbourne
—— and Powell, J.M. (1972) *Phenomenology and Related Non-positivistic Viewpoints in the Social Sciences*, Monash Publications in Geography No. 1, Melbourne
Miliband, R. (1969) *The State in Capitalist Society*, Weidenfeld and Nicolson, London
Mills, C.W. (1959) *The Sociological Imagination*, Oxford University Press, New York
Mingione, E. (1977) 'Theoretical Elements for a Marxist Analysis of Urban Development' in M. Harloe (ed.), *Captive Cities: Studies in the Political Economy of Cities and Regions*, Wiley, Chichester, pp. 89-109
Monstad, M. (1974) 'François Perroux's Theory of "Growth Pole" and "Development" Pole: a Critique', *Antipode*, 6(2), 106-13
Morisita, M. (1959) 'Measuring of the Dispersion of Individuals and Analysis of the Distributional Patterns', *Memoirs, Faculty of Science, Kyushu University*, Series E(Biology), 2, 215-33
Morrill, R.L. (1969) 'Geography and the Transformation of Society: I', *Antipode*, 1(1), 6-9
—— (1970) 'Geography and the Transformation of Society: II', *Antipode*, 2(1), 4-10

—— (1973) 'Socialism, Private Property, the Ghetto and Geographic Theory', *Antipode* 5(2), 84-6

—— (1974) 'Review of D. Harvey (1973), *Social Justice and the City*', *Annals, AAG*, 64, 475-7

—— and Wohlenberg, E.H. (1971) *The Geography of Poverty in the United States*, McGraw-Hill, New York

Moss, R.P. (1979) 'On Geography as Science', *Geoforum*, 10, 223-33

Muth, R. (1969) *Cities and Housing*, University of Chicago Press, Chicago

Nagel, E. (1961) *The Structure of Science: Problems in the Logic of Scientific Explanation*, Harcourt, Brace and World, New York

Neyman, J. and Scott, E.L. (1957) 'On a Mathematical Theory of Populations Conceived as Conglomerations of Clusters', *Cold Spring Harbour Symposia on Quantitative Biology*, 22 (Population Studies), pp. 109-20

Nourse, H.O. and Phares, D. (1975) 'Socioeconomic Transition and Housing Values: a Comparative Analysis of Urban Neighbourhoods' in G. Gappert and H.M. Rose (eds), *The Social Economy of Cities*, Urban Affairs Annual No. 9, Sage Publications, Beverly Hills, pp. 183-208

O'Connor, J. (1973) *The Fiscal Crisis of the State*, St. Martin's, New York

O'Keefe, P. (1979) 'Editorial', *Antipode*, 11(3), 1-2

Ollman, B. (1971) *Alienation: Marx's Conception of Man in Capitalist Society*, Cambridge University Press, London

—— (1972) 'Marxism and Political Science: Prolegomenon to a Debate on Marx's Method', unpublished manuscript, Department of Political Science, New York University, New York

—— (1973) 'Marxism and Political Science: Prolegomenon to a Debate on Marx's Method', *Politics and Society*, 3, 490-508

Olsson, G. (1965) *Distance and Human Interaction: a Review and Bibliography*, Regional Science Research Institute, Bibliography Series No. 2, Philadelphia

—— (1970) 'Logic and Social Engineering', *Geographical Analysis*, 2, 361-75

—— (1972) 'On Reason and Reasoning, on Problems as Solutions and Solutions as Problems, but Mostly on the Silver-tongued Devil and I', *Antipode*, 4(2), 26-31

—— (1974) 'The Dialectics of Spatial Analysis', *Antipode*, 6(3), 50-62

—— (1975) *Birds in Egg*, Michigan Geographical Publications No. 15, University of Michigan

—— (1980) *Birds in Egg/Eggs in Bird*, Pion, London

Park, R.E. (1926) 'The Urban Community as a Spatial Pattern and a Moral Order' in E.W. Burgess (ed.), *The Urban Community*, University of Chicago Press, Chicago

——, Burgess, E.W. and McKenzie, R.D. (1925) *The City*, University of Chicago Press, Chicago

Parsons, T. (1949) *The Structure of Social Action*, Free Press, Glencoe

Paterson, J.L. (1976) 'An Introduction to Philosophical Implications in Geographical Research', BA(Hons) dissertation, Department of Geography, University of Otago, Dunedin

—— (1977) 'Shaping and Informing: Some Thoughts on the Relationship between Geography and Philosophy', *Otago Geographer*, 8, 3-10

Peel, R. (1975) 'The Department of Geography, University of Bristol, 1925-75' in R. Peel, M. Chisholm and P. Haggett (eds), *Processes in Physical and Human Geography: Bristol Essays*, Heinemann, London, pp. 411-17

Peet, R. (1975a) 'The Geography of Crime: a Political Critique', *Professional Geographer*, 27, 277-80 and 28, 96-100

—— (1975b) 'Rural Inequality and Regional Planning', *Antipode*, 7(3), 10-24

—— (1975c) 'Inequality and Poverty: a Marxist-geographic Theory', *Annals, AAG*, 65, 564-71

—— (ed.) (1977a) *Radical Geography: Alternative Viewpoints on Contemporary Social Issues*, Methuen, London

—— (1977b) 'The Development of Radical Geography in the United States', *Progress in Human Geography*, 1, 240-63

—— (1979) 'The Geography of Human Liberation', *Antipode*, 10(3)/11(1), 119-34

—— (1980) 'The Consciousness Dimension of Fiji's Integration into World Capitalism', *Pacific Viewpoint*, 21, 91-115

Philbrick, A.K. (1957) 'Principles of Areal Functional Organization in Regional Human Geography', *Economic Geography*, 33, 299-336

Piaget, J. (1970) *Structuralism*, Basic Books, New York

—— (1972a) *The Principles of Genetic Epistemology*, Routledge and Kegan Paul, London

—— (1972b) *Insights and Illusions of Philosophy*, New American Library, New York

Pickvance, C.G. (ed.) (1976) *Urban Sociology: Critical Essays*, Tavistock, London

Pocock, D.C.D. (1981) 'Place and the Novelist', *Transactions, IBG*, 6, 337-47

Polanyi, K. (1944) *The Great Transformation*, Beacon, Boston

—— (1968) *Primitive, Archaic and Modern Economics: Essays of Karl Polanyi*, edited by G. Dalton, Doubleday, Boston

Popper, K. (1957) *The Poverty of Historicism*, Routledge and Kegan Paul, London

—— (1963) *Conjectures and Refutations*, Routledge and Kegan Paul, London

—— (1968) *The Logic of Scientific Discovery*, Harper and Row, New York

—— (1970) 'Normal Science and its Dangers' in I. Lakatos and A. Musgrave (eds), *Criticism and the Growth of Knowledge*, Cambridge University Press, London, pp. 51-8

—— (1972a) *Objective Knowledge: an Evolutionary Approach*, Clarendon, Oxford

—— (1972b) *Conjectures and Refutations*, 4th edn, revised, Routledge and Kegan Paul, London

—— (1974) 'Replies to my Critics' in P.A. Schilpp (ed.), *The Philosophy of Karl Popper*, vol. 1, Open Court, La Salle, Illinois, pp. 961-1197

—— (1976) *Unended Quest: an Intellectual Autobiography*, Collins, Glasgow

Poulantzas, N. (1973) *Political Power and Social Classes*, New Left Books, London

—— (1975) *Classes in Contemporary Capitalism*, New Left Books, London

—— (1976) 'The Capitalist State: a Reply to Miliband and Laclau', *New Left Review*, 95, 63-83

Pred, A. (1967) *Behavior and Location: Foundations For a Geographic and Dynamic Location Theory*, Part 1, Lund Studies in Geography, Series B, No. 27, Royal University of Lund, Sweden

Prince, H. (1971) 'Questions of Social Relevance', *Area*, 3, 150-3

Radcliffe-Brown, A.R. (1952) *Structure and Function in Primitive Society*, Free Press, Glencoe

Radnitzky, G. (1970) *Contemporary Schools of Metascience*, Akademiforlaget, Goteborg

Regan, C. and Walsh, F. (1976) 'Dependence and Underdevelopment: the Case of Mineral Resources and the Irish Republic', *Antipode*, 8(3), 46-59

Relph, E. (1970) 'An Inquiry into the Relations between Phenomenology and Geography', *Canadian Geographer*, 14, 193-201

—— (1976a) *Place and Placelessness*, Pion, London

—— (1976b) *The Phenomenological Foundations of Geography*, University of Toronto, Department of Geography, Discussion Paper No. 21

—— (1979) 'To See with the Soul of the Eye', *Landscape*, 23, 28-34

Robey, D. (1973) 'Introduction' in D. Robey (ed.), *Structuralism: an Introduction*, Clarendon, Oxford

Robson, B.T. (1970) 'Review of D. Harvey (1969), *Explanation in Geography*', *Environment and Planning*, 2, 363-4

Roder, W. (1961) 'Attitudes and Knowledge on the Topeka Flood Plain' in G.F. White (ed.), *Papers on Flood Problems*, University of Chicago, Department of Geography, Research Paper No. 70, pp. 62-83

Rogers, A. (1965) 'A Stochastic Analysis of the Spatial Clustering of Retail Establishments', *Journal of the American Statistical Association*, 60, 1094-103

Rose, H.M (1971) *The Black Ghetto: a Spatial Behavioral Perspective*, McGraw-Hill, New York

—— (ed.) (1972) *Geography of the Ghetto: Perceptions, Problems, and Alternatives*, Perspectives in Geography, vol. 2, Northern Illinois University Press, De Kalb

Roweis, S.T. and Scott, A.J. (1978) 'The Urban Land Question' in K.R. Cox (ed.), *Urbanization and Conflict in Market Societies*, Methuen, London, pp. 38-75

Rowles, G.D. (1978) 'Reflections on Experiential Field Work' in D. Ley and M. Samuels (eds), *Humanistic Geography: Prospects and Problems*, Maaroufa, Chicago, pp. 173-93

Roxby, P.M. (1930) 'The Scope and Aims of Human Geography', *Scottish Geographical Magazine*, 46, 276-99

Runciman, W.G. (1966) *Relative Deprivation and Social Justice*, Routledge and Kegan Paul, London

Saarinen, T.F. (1966) *Perception of the Drought Hazard on the Great Plains*, University of Chicago, Department of Geography, Research Paper No. 106

Salaman, A. (1934) 'Max Weber's Methodology', *Social Research*, 1

Samuels, M. (1971) 'Science and Geography: an Existential Appraisal', PhD dissertation, Department of Geography, University of Washington

—— (1978) 'Existentialism and Human Geography' in D. Ley and M. Samuels (eds), *Humanistic Geography: Prospects and Problems*, Maaroufa, Chicago, pp. 22-40

—— (1981) 'An Existential Geography' in M.E. Harvey and B.P. Holly (eds), *Themes in Geographic Thought*, Croom Helm, London, pp. 115-32

Santos, M. (1977) 'Spatial Dialectics: the Two Circuits of Urban Economy in Underdeveloped Countries', *Antipode*, 9(3), 49-60

Sauer, C.O. (1925) *The Morphology of Landscape*, University of California Publications in Geography, vol. 2, No. 2, 19-54

—— (1941) 'Foreword to Historical Geography', *Annals, AAG*, 31, 1-24

—— (1952) *Agricultural Origins and Dispersals*, American Geographical Society, New York

Saushkin, Y.G. (1975) 'Review of D. Harvey (1969), *Explanation in Geography*', *Soviet Geography: Review and Translation*, 16, 538-46

Sayer, R.A. (1976) 'A Critique of Urban Modelling: from Regional Science to Urban and Regional Political Economy', *Progress in Planning*, 6, 187-254

—— (1979) 'Epistemology and Conceptions of People and Nature in Geography', *Geoforum*, 10, 19-43

Schaefer, F.K. (1953) 'Exceptionalism in Geography: a Methodological Examination', *Annals, AAG*, 43, 226-49

Schuurman, E. (1977) *Reflections on the Technological Society*, Wedge, Toronto

Seamon, D. (1975) 'The Phenomenological Investigation of Lived Space', *Monadnock*, 49, 38-45

—— (1979a) *A Geography of the Lifeworld: Movement, Rest and Encounter*, Croom Helm, London

—— (1979b) 'Phenomenology, Geography and Geographical Education', *Journal of Geography in Higher Education*, 3, 40-50

Shannon, G.W. and Dever, G.E.A. (1974) *Health Care Delivery: Spatial Perspectives*, McGraw-Hill, New York

Shapere, D. (1964) 'The Structure of Scientific Revolutions', *Philosophical Review*, 73, 383-94

Sherrard, T.D. (ed.) (1968) *Social Welfare and Urban Problems*, Columbia University Press, New York

Simon, H.A. (1957) *Models of Man: Social and Rational*, Wiley, New York

—— (1966) 'Theories of Decision Making in Economics and Behavioural Science', *Surveys of Economic Theory*, vol. 3, 1-28, Macmillan, London

Simon, R.M. (1980) 'The Labour Process and Uneven Development: the Appalachian Coalfields, 1880-1930', *International Journal of Urban and Regional Research*, 4, 46-71

Skellam, J.G. (1958) 'On the Derivation and Applicability of Neyman's Type A Distribution', *Biometrika*, 45, 32-6

Slater, D. (1976) 'The Poverty of Modern Geographical Enquiry', *Pacific Viewpoint*, 16, 159-76

Smith, A. (1970) *The Wealth of Nations*, Penguin, Baltimore

Smith, C.T. (1965) 'Historical Geography: Current Trends and Prospects' in R.J. Chorley and P. Haggett (eds), *Frontiers in Geographical Teaching*, Methuen, London, pp. 118-43

—— (1967) *An Historical Geography of Western Europe before 1800*, Longmans, London

Smith, D.M. (1971) 'Radical Geography – the Next Revolution?', *Area*, 3, 153-7

—— (1973a) *The Geography of Social Well-being in the United States*, McGraw-Hill, New York

—— (1973b) 'Alternative "Relevant" Professional Roles', *Area*, 5, 1-4

—— (1977) *Human Geography: a Welfare Approach*, Edward Arnold, London

Smith, N. (1979) 'Geography, Science and Post-positivist Modes of Explanation', *Progress in Human Geography*, 3, 356-83

Sopher, D.E. (1971) 'Review of D. Harvey (1969), *Explanation in Geography*', *Journal of Regional Science*, 11, 124-7

Spate, O.H.K. (1953) 'The Compass of Geography', Inaugural Lecture, Australian National University, Canberra

Speth, W.W. (1972) 'Historicist Anthropogeography: Environment and Culture in American Anthropological Thought from 1890 to 1950', PhD dissertation, University of Oregon

Spiegelberg, H. (1960) *The Phenomenological Movement: a Historical Introduction*, 2 vols, Martinus Nijhoff, The Hague

Stevens, S.S. (1935) 'The Operational Basis of Psychology', *American Journal of Psychology*, 47, 323-30

Stoddart, D.R. (1981) 'The Paradigm Concept and the History of Geography' in D.R. Stoddart (ed.), *Geography, Ideology and Social Concern*, Basil Blackwell, Oxford, pp. 70-80

Suppe, F. (1974) 'The Search for Philosophic Understanding of Scientific Theories' in F. Suppe (ed.), *The Structure of Scientific Theories*, University of Illinois Press, Urbana, pp. 1-241

Taylor, P.J. (1976) 'An Interpretation of the Quantification Debate in British Geography', *Transactions, IBG*, 1, 129-42

Thackray, A. and Merton, R.K. (1975) 'Sarton, George Alfred Léon' in *Dictionary of Scientific Biography*, vol. 12, Scribner's, New York, pp. 107-14

Thomas, R.W. and Huggett, R.J. (1980) *Modelling in Geography: a Mathematical Approach*, Harper and Row, London

Toulmin, S. (1953) *The Philosophy of Science*, Hutchinson, London

—— (1960) *The Philosophy of Science*, Harper, New York

—— (1970) 'Does the Distinction between Normal and Revolutionary Science Hold Water?' in I. Lakatos and A. Musgrave (eds), *Criticism and the Growth of Knowledge*, Cambridge University Press, London, pp. 39-47

—— (1972) *Human Understanding*, Clarendon, Oxford

Tuan, Yi-Fu (1971) 'Geography, Phenomenology and the Study of Human Nature', *Canadian Geographer*, 15, pp. 181-92

—— (1972) 'Structuralism, Existentialism and Environmental Perception', *Environment and Behavior*, 3, 319-31

—— (1974) 'Space and Place: Humanistic Perspective', *Progress in Geography*, 6, 211-52

—— (1975) 'Place: an Experiential Perspective', *Geographical Review*, 65, 151-65

—— (1976) 'Humanistic Geography', *Annals, AAG*, 66, 266-76

—— (1977a) *Space and Place: the Perspective of Experience*, University of Minnesota Press, Minneapolis

—— (1977b) 'Comment in Reply', *Annals, AAG*, 67, 179-80

—— (1978) 'Literature and Geography: Implications for Geographical Research' in D. Ley and M. Samuels (eds), *Humanistic Geography: Prospects and Problems*, Maaroufa, Chicago, pp. 194-206

Ullman, E.L. (1953) 'Human Geography and Area Research', *Annals, AAG*, 43, 54-66

—— (1956) 'The Role of Transportation' in W.L. Thomas (ed.), *Man's Role in Changing the Face of the Earth*, University of Chicago Press, Chicago, pp. 862-80

Van der Hoeven, J. (1976) *Karl Marx: the Roots of his Thought*, Wedge, Toronto

Van Paassen, C. (1957) *The Classical Tradition of Geography*, J.B. Wolters, Groningen

—— (1976) 'Human Geography in Terms of Existential Anthropology', *Tijdschrift voor Economische en Sociale Geografie*, 67, 324-41

Walker, R.A. (1974) 'Urban Ground Rent: Building a New Conceptual Framework', *Antipode*, 6(1), 51-8

—— (1975) 'Contentious Issues in Marxian Value and Rent Theory: a Second and Longer Look', *Antipode*, 7(1), 31-53

—— (1977) 'The Suburban Solution: Urban Reform and Urban Geography in the Capitalist Development of the United States', PhD dissertation, Department of Geography and Environmental Engineering, Johns Hopkins University, Baltimore

—— (1978) 'Two Sources of Uneven Development under Advanced Capitalism: Spatial Differentiation and Capital Mobility', *Review of Radical Political Economics*, 10(3), pp. 28-34

—— and Storper, M. (1981) 'Capital and Industrial Location', *Progress in Human Geography*, 5, 473-509

Walmsley, D.J. (1974) 'Positivism and Phenomenology in Human Geography', *Canadian Geographer*, 18, 95-107

Warntz, W. and Wolff, P. (1971) *Breakthroughs in Geography*, New American Library, New York

Weber, A. (1928) *Alfred Weber's Theory of the Location of Industries*, Cambridge University Press, London

Weber, M. (1949) *The Methodology of the Social Sciences*, Free Press, Glencoe

Wheatley, P. (1971) *The Pivot of the Four Quarters: a Preliminary Enquiry into the Origins and Character of the Ancient Chinese City*, Aldine, Chicago

White, G.F. (1945) *Human Adjustment to Floods*, University of Chicago, Department of Geography, Research Paper No. 29

Whitehand, J.W.R. (1970) 'Innovation Diffusion as an Academic Discipline: the Case of the "New" Geography', *Area*, 2, 19-30

—— (1971) 'In-words Outwards: the Diffusion of the New Geography', *Area*, 3, 158-63

Whittlesey, D. (1929) 'Sequent Occupance', *Annals, AAG*, 19, 162-5

—— (1945) 'The Horizon of Geography', *Annals, AAG*, 35, 1-36

Whyatt, A. (1978) 'Cooperatives, Women and Political Practice', *International Journal of Urban and Regional Research*, 2, 538-57

Wilson, A.G. (1976a) 'Catastrophe Theory and Urban Modelling: an Application to Modal Choice', *Environment and Planning A*, 8, 351-6

—— (1976b) 'Retailers' Profits and Consumers' Welfare in a Spatial Interaction Shopping Model' in I. Masser (ed.), *Theory and Practice in Regional Science*, Pion, London, pp. 42-57

Winch, P. (1958) *The Idea of a Social Science*, Routledge and Kegan Paul, London

Wisner, B. (1969) 'Editor's Note', *Antipode*, 1(1), iii

Wittgenstein, L. (1961) *Tractatus Logico-philosophicus*, Routledge and Kegan Paul, London

Wolf, L.G. (1976) 'National Economic Planning, a New Economic Policy for America', *Antipode*, 8(2), 64-74

Wolpert, J. (1964) 'The Decision Process in Spatial Context', *Annals, AAG*, 54, 337-58

—— (1965) 'Behavioral Aspects of the Decision to Migrate', *Papers of the Regional Science Association*, 15, 159-69

—— (1971) 'Review of D. Harvey (1969), *Explanation in Geography*', *Annals, AAG*, 61, 180-1

Workman, R.W. (1964) 'What Makes an Explanation', *Philosophy of Science*, 31, 241-54

Yates, E.M. (1971) 'Review of D. Harvey (1969), *Explanation in Geography*', *Geoforum*, 8, 65

Zelinsky, W. (1971) 'Review of D. Harvey (1969), *Explanation in Geography*', *Professional Geographer*, 23, 75-6

—— (1975) 'The Demigod's Dilemma', *Annals, AAG*, 65, 123-43

Zetterberg, H. (1965) *On Theory and Verification in Sociology*, 3rd edn, Bedminster Press, Totawa, New Jersey

# AUTHOR INDEX

# SUBJECT INDEX

Milton Keynes UK
Ingram Content Group UK Ltd.
UKHW031148141024
449569UK00024B/974